Progress in Molecular and Subcellular Biology

27

Springer

Berlin
Heidelberg
New York
Barcelona
Hong Kong
London
Milan
Paris
Singapore
Tokyo

Robert E. Rhoads (Ed.)

Signaling Pathways for Translation

Stress, Calcium, and Rapamycin

With 27 Figures

Springer

Professor Dr. ROBERT E. RHOADS
Department of Biochemistry and Molecular Biology
Louisiana State University Health Sciences Center
Shreveport, LA 71130
USA

ISSN 0079-6484
ISBN 3-540-41710-9 Springer-Verlag Berlin Heidelberg New York

Library of Congress Cataloging-in-Publication Data

Signaling pathways for translation: stress, calcium, and rapamycin / Robert E. Rhoads (ed.).
p. cm. – (Progress in molecular and subcellular biology; 27)
Includes bibliographical references and index.
ISBN 3540417109
1. Proteins – Synthesis. 2. Proteins – Metabolism – Regulation. 3. Cellular signal transduction. 4. Genetic translation. 5. Apoptosis. 6. Calcium – Physiological effect. 7. Stress (Physiology) I. Rhoads, Robert E., 1944– II. Series.
QP551 .S558 2001
572′.645 – dc21

Springer-Verlag Berlin Heidelberg New York
a member of BertelsmannSpringer Science+Business Media GmbH
http.//www.springer.de

Printed in Germany

Cover design: Meta Design, Berlin
Typesetting: Best-set Typesetter Ltd., Hong Kong
SPIN 10797510 39/3130 – 5 4 3 2 1 0 – Printed on acid-free paper

Preface

A diversity of stressful conditions cause rapid and severe inhibition of protein synthesis in eukaryotic cells, in some cases resulting in cell death by apoptosis (programmed cell death). Apoptosis has come to be recognised as an important physiological process in the regulation of growth, development and differentiation. Until recently, relatively little attention had been paid to the changes in protein synthesis during the various phases of apoptosis. The articles in the present volume are by major contributors to our understanding of signaling pathways that result in the inhibition of protein synthesis. These authors trace the downstream consequences of such stress conditions as virus infection, heat shock, nutrient starvation, release of intracellular calcium ions, and treatment with the immunosuppressant rapamycin.

Understanding the mechanisms by which the biosynthesis of proteins is regulated is important for several reasons. Protein synthesis consumes a major portion of the cellular ATP that is generated. Therefore, small changes in protein synthesis can have great consequences for cellular energy metabolism. Translation is also a major site for control of gene expression, since messenger RNAs differ widely in translational efficiency, and changes to the protein synthesis machinery can differentially affect recruitment of individual mRNAs. From a medical standpoint, there is a close but as yet incompletely understood relationship between cell growth and protein synthesis. Understanding the mechanisms regulating protein synthesis therefore holds promise for treatment of diseases that are characterized by uncontrolled cell proliferation as well as those involving tissue degeneration.

The field of cellular signal transduction may be traced back as early as Earl Sutherland's seminal discovery of 3′, 5′-cyclic AMP as a "second messenger", but it has rapidly expanded over the past two decades. The area now includes cell surface receptors and a host of intermediate components that operate in branching, and often interconnecting, pathways to transmit an extracellular signal to the ultimate cellular targets. Thus, an encoded chemical message is translated into the language of the cell by altering a biochemical activity. The complexity of the networks thus far elucidated permits multiple simultaneous signals to be integrated and also allows a single signal to modulate a variety of target activities. Differences in signaling pathways in various cell types allows for cell-specific responses to the same agonist.

The process of protein synthesis is itself one of the most complex biochemical achievements of living cells, resulting in the translation of an encoded genetic message into proteins that carry out both structural and functional roles. This rapid and highly accurate process requires the coordinated action of roughly 150 different polypeptides and 70 different RNAs. Discovering how extracellular signals might regulate such an intricate and complex process represents an enormous challenge.

Despite such complexity, considerable progress has been made regarding this important question. Surprisingly, only a few of the many components of the protein synthesis machinery have been implicated as targets of signaling pathways. The majority of these involve individual polypeptide components of the initiation and elongation factors. The change in activity of the factor is generally accompanied by phosphorylation of the factor or phosphorylation of a binding partner, changing either the intrinsic activity of the factor or the affinity of its association with the binding partner. In several cases, the end result of a signaling cascade is proteolysis of the factor.

Given the subtlety and complexity of both the network of signaling pathways and the machinery for protein synthesis, it is likely that our current knowledge represents merely "the tip of the iceberg". It is the authors desire that the current work will stimulate further research in this interesting and important field.

Shreveport, Louisiana, USA
May 2001

ROBERT E. RHOADS

Contents

Regulation of the Activity of Eukaryotic Initiation Factors in Stressed Cells
Gert C. Scheper, Roel Van Wijk, and Adri A.M. Thomas

Initiation Factor eIF2α Phosphorylation in Stress Responses and Apoptosis
Michael J. Clemens

Elongation Factor-2 Phosphorylation and the Regulation of Protein Synthesis by Calcium
Angus C. Nairn, Masayuki Matsushita, Kent Nastiuk, Atsuko Horiuchi, Ken-Ichi Mitsui, Yoshio Shimizu, and H. Clive Palfrey

The Regulation of eIF4F During Cell Growth and Cell Death

Simon J. Morley[1]

1 Introduction

Much effort has been focused on questions concerning the highly regulated processes of cell growth, proliferation and programmed cell death (apoptosis). Hormones, growth factors and ligands exert pleiotypic effects through activation of specific cell-surface receptors, and via transmembrane signalling and activation of common protein kinase/phosphatase cascades inside the cell. These, in turn, trigger an array of cellular responses, culminating in either cell growth and division, differentiation or cell death. One of the obligatory, early responses in all of these processes is a modulation of the rate of protein synthesis, mediated by regulating the phosphorylation of translation initiation factor polypeptides and their association into functional complexes. This chapter will review current knowledge about the regulation of these initiation factor proteins in response to cell growth and cell death.

2 Mechanism and Regulation of Initiation of Protein Synthesis

Protein synthesis involves a complex series of protein:protein and protein: RNA interactions culminating in the formation of peptide bonds between amino acids, as encoded by the mRNA being translated (reviewed in Merrick and Hershey 1996; Pain 1996; Gingras et al. 1999b). This process, summarised in Fig. 1, is described in more detail below. Briefly, the initiation phase of protein synthesis can be considered as four successive events: (1) the formation of a ternary complex between initiation factor eIF2, GTP and initiator methionyl tRNA (Met-tRNA$_f$); (2) binding of the ternary complex to the 40 S ribosomal subunit to form the 43 S initiation complex; (3) binding of mRNA to the 43 S initiation complex; (4) coupling of the 60 S ribosomal subunit to the mRNA/ribosome complex, release of bound initiation factors and formation of the 80 S initiation complex. Each stage involves the interaction of several

[1] Biochemistry Laboratory, School of Biological Sciences, University of Sussex, Falmer, Brighton BN1 9QG, UK

Progress in Molecular and Subcellular Biology, Vol. 27
Robert E. Rhoads (Ed.)

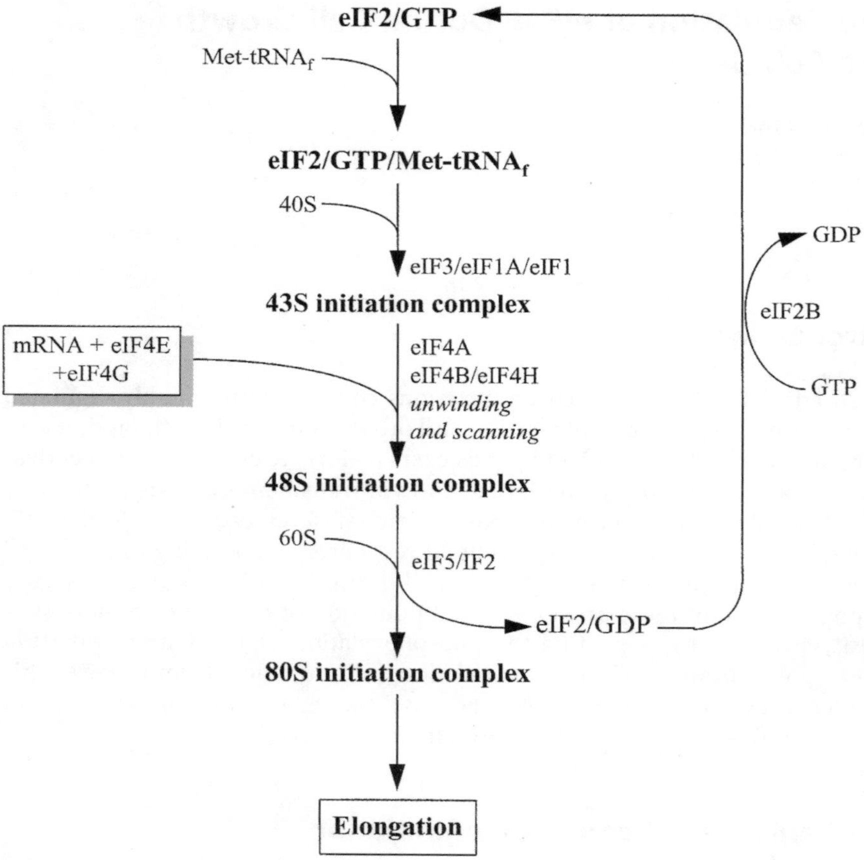

Fig. 1. Summary of the initiation of protein synthesis. A complex of the initiation factor eIF2 with GTP binds the initiator Met-tRNA$_f$ to form a [Met-tRNA$_f$/eIF2/GTP] ternary complex. This then associates with a 40 S ribosomal subunit carrying eIF3 (and other factors such as eIF1/eIF1 A) to produce the 43 S preinitiation complex. mRNA binding to this complex involves the participation of the trimeric eIF4F complex, comprising the cap-binding protein eIF4E, the scaffold protein eIF4G and the RNA helicase eIF4A, together with the helicase-stimulatory factor eIF4B/eIF4H (see text for details). In the final common step, the anticodon of Met-tRNA$_f$ interacts with the AUG codon, facilitated by eIF2 and eIF4B, to generate the mature 48 S initiation complex. Following the location of the AUG initiation codon on the mRNA with the aid of the eIF1 group of proteins (Pestova et al. 1998, 2000), two additional events must occur. The GTP associated with the eIF2 is hydrolysed to GDP, in parallel with the dissociation of the previously bound initiation factors from the ribosome (Merrick and Hershey 1996; Dever 1999; Gingras et al. 1999). GTP hydrolysis requires the involvement of the eIF5 group of initiation factors and a homologue of bacterial IF2 (Pestova et al. 2000). The GDP that is generated remains associated with the eIF2 and must be exchanged for another molecule of GTP in a process catalysed by the guanine nucleotide exchange factor eIF2B (see C.G. Proud, Vol. I). The second event is the binding of the larger (60 S) ribosomal subunit to the initiation complex to form the complete 80 S initiation complex. (Modified from Merrick 1992; Pain 1996)

components, the majority of which have been purified to homogeneity and utilised to study the individual reactions in vitro. In the best characterised systems, control of protein synthesis is exerted at the level of initiation, although there is some evidence for regulation at the level of elongation (reviewed in Merrick and Hershey 1996; Pain 1996; Sonenberg 1996; Proud and Denton 1998; Kleijn et al. 1998; ; Dever 1999; Gingras et al. 1999b; Preiss and Hentze 1999; Rhoads 1999).

Two stages in the initiation phase have been found to be highly susceptible to regulation by physiological conditions, with the phosphorylation of the factors involved playing a key regulatory role. The first of these, the binding of the eIF2/GTP/Met-tRNA$_f$ (ternary) complex to the 43 S ribosomal subunit, is limited by the recycling of eIF2 and regulated through phosphorylation of eIF2α and the physiological control of initiation factor eIF2B (Fig. 1). This has been shown to be a key control point in the acute down regulation of protein synthesis in response to physiological stress or metabolic challenges and is discussed elsewhere in this volume. The second stage, which is the focus of this chapter, is the association of this pre-initiation complex with an mRNA molecule (Fig. 2). This event is mediated by initiation factors of the eIF4 family which act in concert to find the 5′ end of the mRNA, unwind any secondary structure therein, direct the binding of the 43 S initiation complex and, along with further unwinding of the 5′ untranslated region, allow migration of the complex to the initiation codon. As such, the eIF4 initiation factors are responsible for recruiting mRNA to the ribosome in a mechanism with strong analogies to transcription complex assembly at promoter sites on DNA (Morley 1996; Sachs and Buratowski 1997).

2.1
The mRNA-Binding Initiation Factors

2.1.1
Initiation Factor eIF4E

All cytoplasmic mRNAs have a unique cap structure at their 5′ terminus, with the general composition m^{7}G(5′)ppp(5′)N (where *N* is any nucleotide). The presence of this cap structure has a strong stimulatory effect on the translation of mRNA, with uncapped or nonmethylated mRNAs being less competent at 48 S and 80 S initiation complex formation (Sonenberg 1996). Several cytoplasmic proteins can specifically interact with the mRNA cap structure; as judged by cross-linking, four initiation factors, eIF4E, eIF4A, eIF4B and eIF4G have been defined. Of these, only the phosphoprotein eIF4E specifically interacts directly with the cap, either as an individual polypeptide or part of a protein complex, termed eIF4F (Fig. 2). Mutagenesis experiments on the eIF4E protein have suggested that highly conserved tryptophan residues are important for interaction with the cap structure (Marcotrigiano et al. 1997; Altmann et al. 1988). These findings have been confirmed and explained at the bio-

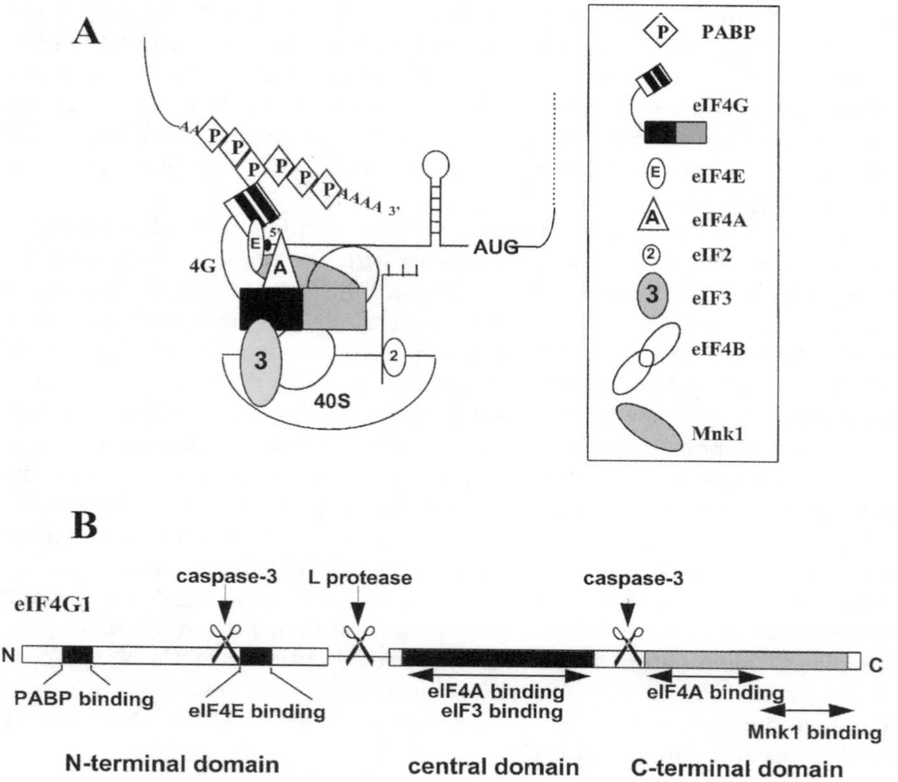

Fig. 2A,B. A current model for 48 S initiation complex formation in mammalian cells. **A** Interaction between individual initiation factors, the protein kinase, Mnk1, and the 5′ and 3′ ends of the mRNA (see text for details). **B** Diagrammatic representation of mammalian eIF4GI, showing known sites of interaction with PABP, eIF4E, eIF3, eIF4A and Mnk1 (Lamphear et al. 1995; Mader et al. 1995; Imataka et al. 1997, 1998; Morley et al. 1997; Piron et al. 1998; Gradi et al. 1998a; Gingras et al. 1999b; Pyronnet et al. 1999; Morino et al. 2000). Also indicated are the sites of cleavage by picornavirus L protease (Kirchweger et al. 1994) and caspase-3. (Bushell et al. 2000b)

chemical level by the solving of the three dimensional structure of mouse eIF4E. These structural studies showed that eIF4E resembles a cupped hand, with the m^7GTP cap occupying a narrow slot on the concave side and interacting proteins binding to separate, but overlapping, sites on the convex surface (Marcotrigiano et al. 1997, 1999; Gingras et al. 1999b; Ptushkina et al. 1999; Sonenberg, this Vol.). Forced overexpression of eIF4E results in transformation to a malignant phenotype (De Benedetti and Rhoads 1990; Lazaris-Karatzas et al. 1990). It has been suggested that this may reflect the enhanced translation of "inefficient" mRNAs (Koromilas et al. 1992), such as those encoding signalling molecules (Marth et al. 1988), ornithine decarboxylase (Rousseau et al. 1996), cyclin D1 (Rosenwald et al. 1995) and ribonucleotide reductase (Abid et al. 1999). Moreover, raised levels of expression of eIF4E have been described in various human tumours (Li et al. 1997; Nathan et al. 1997b; Rosenwald et al. 1999), while down regulation of eIF4E strongly inhibits protein synthesis and cell growth (Rinker-Schaeffer et al. 1993).

2.1.2
eIF4E Phosphorylation

Although the site of phosphorylation of eIF4E was initially identified as Ser53 (Rychlik et al. 1987) and numerous studies were made with Ser53Ala mutant proteins (De Benedetti and Rhoads 1990; Joshi-Barve et al. 1990; Sonenberg 1994, 1996), it is now clear that phosphorylation of eIF4E occurs at Ser209 (Joshi et al. 1995; Flynn and Proud 1995; Whalen et al. 1996). To date, the effects of Ser209 mutations on cell growth have not been reported. As correlations exist between the enhanced phosphorylation of eIF4E and increased rates of protein synthesis (reviewed in Rhoads 1993; Morley 1996; Sonenberg 1996; Kleijn et al. 1998; Gingras et al. 1999b), important questions remain on the functional consequences of eIF4E phosphorylation and the signal transduction pathways which mediate it. While phosphorylation of eIF4E has been reported to increase its interaction with the cap structure (Minich et al. 1994), and the association of eIF4E with eIF4G enhances the binding of eIF4E to the mRNA cap (Haghighat and Sonenberg 1997), the precise mechanistic consequences of eIF4E phosphorylation are unclear. We (Morley 1997; Morley and McKendrick 1997; Fraser et al. 1999a, b), and others (Flynn and Proud 1996b; Wang et al. 1998; Gingras et al. 1999b) have shown that phosphorylation of eIF4E occurs via multiple signalling pathways. Indeed, recent studies have shown that the protein kinase Mnk1, which acts at the convergence point of mitogen-activated protein kinase (MAPK/ERK) and p38 MAP kinases (Waskiewicz et al. 1997; Wang et al. 1998), phosphorylates eIF4E at the physiological site in vitro and in vivo (Pyronnet et al. 1999; Waskiewicz et al. 1999). The related kinase, Mnk2, linked exclusively to the ERK pathway, phosphorylates eIF4E less efficiently (Waskiewicz et al. 1997). Both kinases interact directly with eIF4GI and eIF4GII (Pyronnet et al. 1999; Morino et al. 2000) bringing them in close proximity to eIF4E within the eIF4F complex (Fig. 2). However, although expression of a dominant-negative mutant of Mnk1 interfered with the phorbol ester-stimulated phosphorylation of co-transfected eIF4E (Waskiewicz et al. 1999), effects on global or specific translation were not reported.

2.2
Initiation Factor eIF4A

eIF4A shows the properties of RNA-dependent ATP hydrolysis and ATP-dependent RNA binding in vitro (Grifo et al. 1984; Ray et al. 1985; Pause and Sonenberg 1992; Pause et al. 1994b). In conjunction with eIF4B (see below), eIF4A is believed to function to promote unwinding of mRNA secondary structure (Rozen et al. 1990; Pause and Sonenberg 1993; Pause et al. 1994b). It is absolutely required for mRNA-ribosome binding, both in its free form and as part of the eIF4F complex (see below), and is essential for translation of all mRNAs and for growth in yeast (Altmann et al. 1990). eIF4A can be cross-linked to the mRNA cap structure and other initiation factors in the presence of ATP, despite a lack of intrinsic cap binding ability. cDNA sequencing indicated that the protein contains the DEAD box motif present in RNA unwinding proteins (Linder and Slonimski 1989), and a glycine rich region characteristic of ATP and GTP binding proteins (Linder et al. 1989; De la Cruz et al. 1999). There are three highly homologous mammalian isoforms which appear to differ in tissue distribution and developmental regulation (Yoder-Hill et al. 1993; Weinstein et al. 1997; Williams-Hill et al. 1997; De la Cruz et al. 1999; Li et al. 1999). It has been suggested that eIF4A functions primarily as a subunit of the eIF4F complex, with eIF4A required to recycle through the complex during translation (Pause et al. 1994b; Merrick and Hershey 1996; Sonenberg 1996; Li et al. 1999).

2.3 eIF4B and eIF4H

eIF4B, a phosphoprotein, can also be cross-linked to the mRNA cap but only in the presence of eIF4A, eIF4E, eIF4G and ATP (Edery et al. 1983; Grifo et al. 1983), whilst alone it has a weak affinity. This highly phosphorylated, but poorly understood initiation factor appears to function as a dimer, is required for mRNA binding to ribosomes (Benne and Hershey 1978), and stimulates the RNA helicase activity of eIF4A in vitro (Merrick 1992; Merrick and Hershey 1996; Sonenberg 1996). Three potential regulatory domains are revealed by the cDNA sequence: an RNA binding domain (RRM), a hydrophilic region (DRYG), which mediates binding of a truncated form of eIF4B to eIF3p170 (Methot et al. 1997), and a serine rich region at the C-terminus (Milburn et al. 1990). In vitro studies with mutant proteins have indicated that the RNA binding domain alone is insufficient to support interaction with mRNA, but may localise eIF4B to the ribosome (Methot et al. 1994); a region in the DRYG domain, however, is important for both the RNA binding and ability of eIF4B to stimulate the helicase activity of eIF4A (Naranda et al. 1994). Thus, eIF4B has been postulated to act directly by binding to the 5′ UTR of the mRNA and to the ribosome, and/or indirectly via its interaction with eIF3 to promote the mRNA/rRNA/initiator tRNA interaction at the AUG codon. Recently, an eIF4B-related protein (eIF4H) has been identified which possesses an RRM domain but lacks the corresponding DRYG domain (Richter-Cook et al. 1998). eIF4H can substitute for eIF4B in a reconstituted translation system, increases the affinity of eIF4A for RNA, stimulates eIF4A helicase activity (Rogers et al. 1999), and may function to directly stabilise conformational changes in eIF4A that occur during initiation.

2.4 Scaffold Protein eIF4G

eIF4G, of which there are two forms in mammals (eIF4GI, eIF4GII), functions as an adapter protein in the assembly of the initiation complex, eIF4F (Fig. 2). In addition, two families of proteins with sequence homology to eIF4G have been identified, referred to as p97/NAT1/DAP-5 and Paip-1 (reviewed in Morley et al. 1997; Gingras et al. 1999b). Within the sequence of eIF4GI and eIG4GII, there are specific domains that interact with eIF4E (Mader et al. 1995; Gradi et al. 1998a; Imataka et al. 1998), two binding sites for eIF4A (Lamphear et al. 1995; Imataka and Sonenberg 1997; Morino et al. 2000), and binding sites for eIF3 (Lamphear et al. 1995; Morino et al. 2000), RNA (Goyer et al. 1993), poly(A)-binding protein (PABP) (Tarun and Sachs 1996; Tarun et al. 1997; Imataka et al. 1998; Piron et al. 1998) and the eIF4E kinase, Mnk1 (Gingras et al. 1999b; Pyronnet et al. 1999; Morino et al. 2000). Interaction of PABP with eIF4G has been suggested to facilitate the functional association of the 3′ end of an mRNA with the 5′ end and increase the helicase asctivity of eIF4F (Bi and Goss 2000), while the association of eIF4GI with eIF4E markedly enhances the binding of this complex to the mRNA cap (Haghighat and Sonenberg 1997). Increased phosphorylation of eIF4G has been associated with the up-regulation of cell growth in a number of cell systems (Morley and Traugh 1990, 1993; Morley and Pain 1995a, b; Morley et al. 1997; Fraser et al. 1999a) and this appears to be intimately associated with the increase in the phosphorylation of eIF4E (Morley 1996; Kleijn et al. 1998; Gingras et al. 1999b), probably facilitated by the binding of Mnk1 to the C-terminus of eIF4GI (Pyronnet et al. 1999; Morino et al. 2000). Recently, it has been shown that although Mnk1 will phosphorylate eIF4GI in vitro, it is by no means clear which kinase(s) function in vivo. It is, however, known that the activity of the PI3-K/PKB/mTOR and MAP kinase kinase (MEK) pathways is required for the regulated phosphorylation of eIF4G but the functional consequences of phosphorylation on eIF4G activity are unknown at this time (Raught et al. 2000).

2.5 Poly(A)-Binding Protein (PABP)

PABP is found in all eukaryotes, where it plays an essential role for growth in *S. cerevisiae* (Sachs et al. 1987) and may mediate the stimulatory effects of the poly(A) tail on translation in mammalian cells (Gallie and Traugh 1994; Jackson and Wickens 1997; Wickens et al. 1997). In *S. cerevisiae*, the RNA recognition motif 2 (Kessler and Sachs 1998) of Pab1p has been shown to interact with eIF4G at a specific sequence N-terminal to the eIF4E binding site (Tarun and Sachs 1996; Tarun et al. 1997). In vivo (Imataka et al. 1998; Piron et al. 1998), PABP binds directly to a domain in the N-terminus of eIF4G via a sequence of basic amino acids which bears no homology to the yeast sequence (Fig. 2). This association is believed to mediate the circularisation of mRNA and promote the poly(A) and PABP-dependent stimulation of mRNA translation previously demonstrated (Gallie 1996; Jacobson 1996; Le et al. 1997; Sachs and Buratowski 1997; Tarun et al. 1997; Wickens et al. 1997; Gray and Wickens 1998; Wells et al. 1998). In addition, PABP interacts with a protein, Paip-1, which has sequence homology to the middle region of eIF4GI (Craig et al. 1998); however, the role of this protein in translation initiation is not clear.

2.6 eIF4E Binding Proteins

It is believed that effective concentrations of eIF4E can be physiologically modulated on an acute basis by association with a family of regulatory proteins, 4E-BP1, 4E-BP2 and 4E-BP3 (Lin et al. 1994; Pause et al. 1994a; Sonenberg 1996; Lawrence and Abraham 1997; Proud and Denton 1998; Poulin et al. 1998; Gingras et al. 1999a, b). 4E-BP1 was found to be 93% identical to PHAS-I, which had been identified previously as a protein rapidly phosphorylated in adipose tissue in response to insulin (reviewed in Denton and Tavaré 1995; Lawrence and Abraham 1997). Interaction of 4E-BP1 with eIF4E inhibits cap structure-dependent translation both in vitro and when the protein is expressed in cells (Rousseau et al. 1996; Gingras et al. 1999b). In vitro studies showed that 4E-

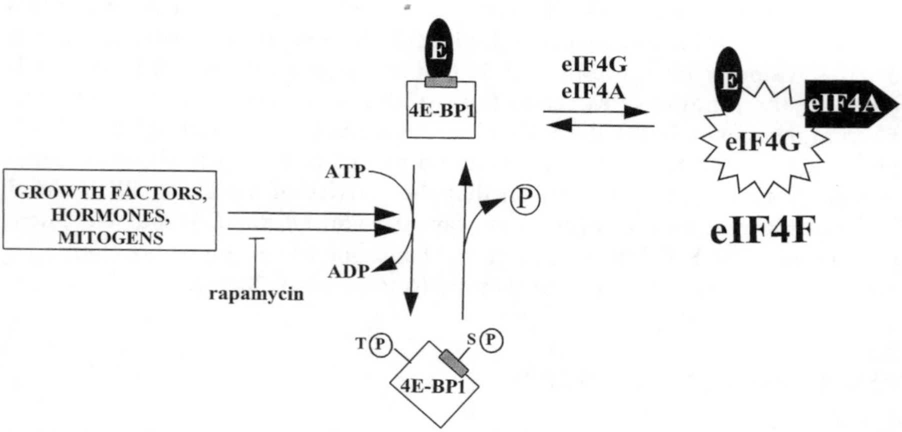

Fig. 3. Proposed role of 4E-BP1 in eIF4F complex assembly. Growth factors interact with their respective receptors at the membrane to activate an array of rapamycin-sensitive and -insensitive intracellular signalling pathways, culminating in the enhanced phosphorylation of 4E-BP1 at defined sites (Lin et al. 1994; Pause et al. 1994a; Diggle et al. 1996; Lawrence and Abraham 1997; Heesom et al. 1998; Gingras et al. 1999a, b; Heesom and Denton 1999; Yang et al. 1999). Phosphorylation of 4E-BP1 on a subset of these sites results in its dissociation from eIF4E allowing eIF4E to interact with the N-terminal domain of eIF4G via a sequence motif conserved between 4E-BP1 and eIF4G (Mader et al. 1995; Gingras et al. 1999b). This allows for eIF4F complex formation and translation to proceed; maintenance of eIF4F complex levels may also reflect the phosphorylation of 4E-BP1 by distinct kinases once it has dissociated from eIF4E (Heesom and Denton 1999) and/or by regulation of protein phosphatase activity

BP1 prevented the binding of mRNA to the ribosome by competing with eIF4G for binding to eIF4E and preventing the assembly of eIF4E into the eIF4F complex (Haghighat et al. 1995; Mader et al. 1995). Conversely, eIF4E already in a complex with eIF4G could not interact with 4E-BP1 (Haghighat et al. 1995; Rau et al. 1996). As depicted in Fig. 3, it is believed that, in vivo, the hormone-induced hyperphosphorylation of 4E-BP1 in a wide variety of cell types causes it to dissociate from eIF4E (Denton and Tavaré 1995; Diggle et al. 1996; Lawrence and Abraham 1997; Heesom et al. 1998; Proud and Denton 1998; Gingras et al. 1999a, b; Heesom and Denton 1999). This allows eIF4E to interact with the N-terminal domain of eIF4G via a conserved motif (Mader et al. 1995; Marcotrigiano et al. 1999; Ptushkina et al. 1999), leading to enhanced levels of the eIF4F complex (reviewed in Morley 1996; Flynn and Proud 1996a; Pain 1996; Sonenberg 1996; Dever 1999; Gingras et al. 1999b; Preiss and Hentze 1999). However, the importance of the eIF4E interaction with the 4E-BPs as a mechanism regulating translation seems to vary considerably between different cell systems (see Sect. 3).

3
The Regulation of eIF4F Complex Levels

3.1
eIF4F: An Overview

The scientific literature contains many examples of general correlations between enhanced rates of eIF4E phosphorylation and an increase in the binding of eIF4E to the scaffold protein, eIF4G, to form the functional initiation factor complex, eIF4F. These are documented as occurring in response to serum, growth factors, hormones, phorbol esters, phosphatase inhibitors, heat shock reversal, meitoic maturation and fertilisation of eggs, as well as during the expression of oncogenic signalling molecules (extensively reviewed in Rhoads 1993; Morley 1994, 1996; Pain 1996; Sonenberg 1996; Kleijn et al. 1998; Proud and Denton 1998; Gingras et al. 1999b; McKendrick et al. 1999; Rhoads 1999). Apart from changes in initiation factor phosphorylation, eIF4F complex formation may be mediated in part by competitive interactions that modulate the ability of eIF4E and eIF4G to form the eIF4F complex (Figs. 2 and 3), and by regulation of the integrity of eIF4G (see below). Additionally, the eIF4F complex can also be found associated with PABP in response to serum refeeding of cultured *Xenopus* kidney cells (Fraser et al. 1999a), effectively resulting in mRNA circularisation, proposed to enhance translational efficiency (Gallie 1996; Jacobson 1996; Wickens et al. 1997; Preiss and Hentze 1999; Bi and Goss 2000). In some, but not all, studies these events have also been associated with a stimulation of general protein synthesis.

It is generally accepted that the eIF4F complex and associated proteins are localised to the 5′ end of mRNA and unwind nearby mRNA secondary structure, allowing interaction of the 43 S initiation complex with the AUG start codon. However, the sequence of events resulting in the assembly on eIF4F on the mRNA cap structure is still a matter of some debate (Morley 1994; Pain 1996; Kleijn et al. 1998; Muckenthaler et al. 1998; Gingras et al. 1999b; Paraskeva et al. 1999). What is obvious at this stage is the lack of knowledge we have regarding the order of association of the initiation factors with each other and with mRNA, as well as the stage at which these factors dissociate from either the mRNA cap or the ribosome during translation. In addition, there are now an increasing number of exceptions to the generally accepted model, indicating that more than one pathway may be operational at any one time. Moreover, although the physiological regulation of eIF4E/4E-BP1 association has been clearly demonstrated, it is only in a few instances that the control of eIF4F complex formation has been clearly connected with observations of parallel, but reciprocal, changes in the degree of association of cellular eIF4E with eIF4G and 4E-BP1.

3.2 eIF4F and Increased Rates of Protein Synthesis

In many cell types, stimulation with growth factors or hormones results in the phosphorylation of both eIF4E and 4E-BP1, with consequent release of the associated eIF4E, allowing the latter to interact with eIF4G and promote the recruitment of mRNA into polysomes (reviewed in Morley 1996; Kleijn et al. 1988; Gingras et al. 1999b). Figure 4A shows a typical result from our laboratory in which these events are correlated with stimulation of protein synthesis rates during serum-refeeding of NIH3T3 cells (Morley and McKendrick 1997). However, these correlations are by no means universal. There are now several reports of physiological situations where protein synthesis rates were raised in parallel with increases in eIF4F complex levels but in the absence of any change in the association of eIF4E with 4E-BP1 (Wada et al. 1996; Yoshizawa et al. 1997; Fraser et al. 1999a, b; Tuxworth et al. 1999; Vary et al. 1999, 2000). Figure 4B shows a result of this type from our laboratory in which serum refeeding of *Xenopus* kidney cells promoted the phosphorylation of eIF4E and the association of eIF4A, eIF4G and PABP with eIF4E, but induced no change in the association of 4E-BP1 or 4E-BP2 with eIF4E (Fraser et al. 1999a). Moreover, the converse situation can also occur, where conditions that promote the dephosphorylation of 4E-BP1 and increase the association of 4E-BP1 with eIF4E failed to influence the rate of protein synthesis and/or eIF4F complex formation (Fig. 4A; Morley and McKendrick 1997; Fraser et al. 1999a; Marx and Marks 1999; Tinton and Buc-Calderon 1999) or cell cycle progression (Terada et al. 1993, 1995).

These data suggest that there may be more than one functional population of eIF4E in the cell, each subject to distinct mechanisms of regulation. This conclusion is strengthened by reports (summarised in Table 1) where correlations of increased translation rates with the association and/or phosphorylation of initiation factors do not fit any one simple model. Furthermore, studies

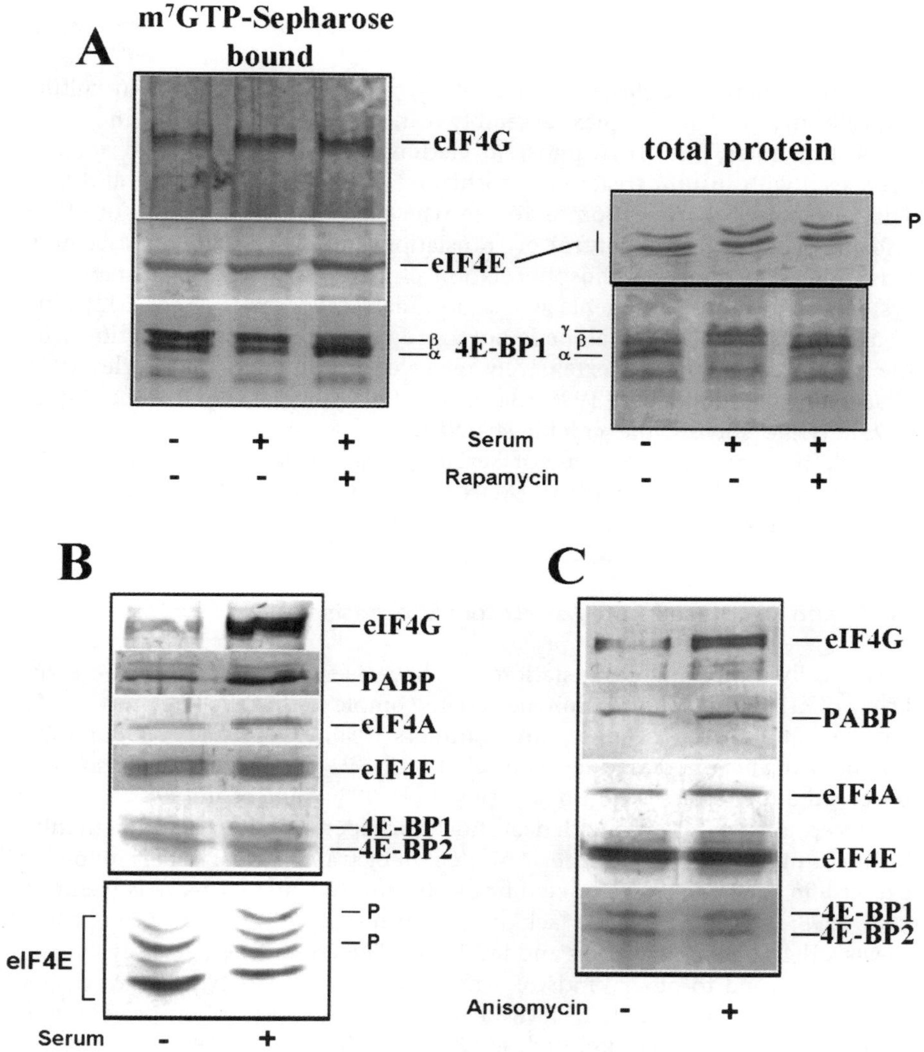

Fig. 4A–C. Increased association of eIF4G with eIF4E is not always associated with a decrease in 4E-BP1/eIF4E complexes. **A** NIH 3T3 cells were serum-starved for 48 h, then incubated for 4 h in the absence or presence of 50 nM rapamycin or 10% serum, as indicated. Aliquots of cell extracts (40 μg) were subjected to m^{7}GTP-Sepharose to isolate eIF4E and associated proteins, prior to analysis by SDS-PAGE and immunoblotting with antiserum specific to 4E-BP1, eIF4E and eIF4G (*left panel*). In addition, the phosphorylation status of eIF4E was visualised by isoelectric focusing (VSIEF) and immunoblotting (*upper right*), and that of 4E-BP1 visualised by SDS-PAGE and immunoblotting (*lower right*). The α-form of 4E-BP1 is the less phosphorylated and the γ-form is the most highly phosphorylated form of 4E-BP1 (data from Morley and McKendrick 1997). **B** Serum-starved *Xenopus* kidney cells were incubated in the absence or presence of 10% serum for 30 min. Cell extracts were prepared, adjusted to equal protein concentration and eIF4E and associated proteins isolated by m^{7}GTP-Sepharose chromatography, prior to analysis by SDS-PAGE and immunoblotting with antiserum specific to 4E-BP1, 4E-BP2, eIF4E, eIF4A, PABP and eIF4G (*upper panel*). In addition, the phosphorylation of eIF4E was visualised by VSIEF and immunoblotting; the more phosphorylated forms of eIF4E are indicated (data from Fraser et al. 1999a). **C** Serum-starved *Xenopus* kidney cells were incubated in the absence or presence of 10 μg/ml anisomycin for 30 min. Cell extracts were prepared and eIF4F complexes were visualised as described in **B**. (Data from Fraser et al. 1999b)

involving skeletal muscle during refeeding in vivo and myoblasts in culture indicate that eIF4F complex assembly can be associated with increased translation rates, but eIF4E phosphorylation was not essential and actually decreased with insulin treatment (Yoshizawa et al. 1997; Kimball et al. 1998). This also raises the question as to the true physiological function of eIF4E phosphorylation in the control of translation and eIF4F complex assembly. It is quite clear that the phosphorylation of eIF4E is increased in many cell systems in response to agents which stimulate cell growth (Sect. 2.1.2), but phosphorylation of eIF4E alone is not a pre-requisite for its interaction with the mRNA cap structure (extensively reviewed in Rhoads 1993; Morley 1994, 1996; Pain 1996; Sonenberg 1996; Kleijn et al. 1998; Gingras et al. 1999b; Rhoads 1999). While solving the crystal structure has generated speculation over possible mechanisms for a role of Ser209 phosphorylation in stabilising the interaction of mRNA with eIF4E (Marcotrigiano et al. 1997), these await biochemical confirmation.

3.3 eIF4F and Decreased Rates of Protein Synthesis

Historically, inhibition of translation rates have been associated with decreased levels of eIF4E phosphorylation and eIF4F complexes in cells (e.g. heat shock; Panniers et al. 1985; Lamphear and Panniers 1990, 1991; Duncan 1996). Consistent with this, increased levels of eIF4E/4E-BP1 complexes have also been observed under stress conditions (Vries et al. 1997; Wang et al. 1998), but there are exceptions to this even with heat shock (Scheper et al. 1997). Although inhibition of muscle protein synthesis by alcohol or translational effects following exercise in skeletal muscle in vivo fit nicely with the current models (Gautsch et al. 1998; Lang et al. 1999), a lack of correlation between rates of protein synthesis, eIF4E phosphorylation and levels of eIF4F complexes is clearly evident in cells exposed to other kinds of stress (Table 1 and Fig. 4C). Paradoxically, under such conditions, eIF4F levels can actually increase when protein synthesis rates are severely inhibited (Morley and McKendrick 1997; Fraser et al. 1999b; Shah et al. 1999). One explanation for these apparently conflicting results could be that increased phosphorylation of eIF4E and enhanced levels of eIF4F complex formation, mediated by stress kinase signalling pathways (Waskiewicz et al. 1997; Wang et al. 1998; Fraser et al. 1999b), play a specific role in promoting de novo recruitment of mRNAs into the translated pool and/or by influencing mRNA stability (Laroia et al. 1999; Winzen et al. 1999). Such effects could be important in promoting cell survival. Further work is required to address this possibility.

Table 1. Changes in initiation factor modification, association and eIF4F complex levels do not always correlate with protein synthesis rates

Cell or tissue	Treatment	eIF4E(P)	4E-BP1(P)	eIF4E/BP1 complexes	eIF4F complexes	Protein synthesis	Reference
Muscle	Refeeding	↔	↑	↓	↑	↑	Yoshizawa et al. (1997)
Muscle	Insulin	↓	↑	↓	↑	↑	Kimball et al. (1997)
Myoblasts	Insulin	↓	↑	↓	↑	↑	Kimball et al. (1998)
Muscle	IGF-1	↔	↔	↔	↑	↑	Vary et al. (2000)
Liver	Refeeding	↑	↔	↔	↑	↑	Yoshizawa et al. (1997)
	Supraphysiological amino acids	↑	↔	↔	↑	↑	Vary et al. (1999)
	Cardiac load	↑	↔	↔	↑	↑	Tuxworth et al. (1999)
Xenopus cells	Serum refeeding	↑	↔	↔	↑	↑	Fraser et al. (1999)
Muscle	Glucocorticoids	↑	↓	↑	↓	↓	Shah et al. (2000)
Liver	Amino acid imbalance	–	↑	↓	↑	↓	Shah et al.(1999)
NIH3T3 cells	Anisomycin	↑	↑	↓	↑	↓	Morley and McKendrick(1997)
Starved *Xenopus* cells	Arsenite, anisomycin, heat shock	↑	↔	↔	↑	↓	Fraser et al. (1999)
Hepatoma cells	Heat shock	↓	↑	↓	–	↓	Scheper et al. (1997)
Cardiac myocytes	Arsenite	↑	↑	↓	–	↔	Wang and Proud (1997)
Human kidney cells	Arsenite	↑	↔	↔	↔	–	Wang et al. (1998)
Human kidney cells	H_2O_2, sorbitol	↔	–	↑	↓	–	Wang et al. (1998)

↑ increase; ↓ decrease; ↔ no change; – not determined.

3.4 Internal Ribosome Binding, eIF4E, eIF4G Integrity and Reduced Levels of eIF4F

3.4.1 Initiation on Viral Internal Ribosome Entry Sites

An alternative mechanism for the recruitment of ribosomes to mRNA is that of internal ribosome binding (Jackson 1995). This process, which is mediated by the direct binding of the 40 S ribosome to an internal RNA structure (IRES) bypassing the 5′ cap, was first reported for picornaviral mRNAs (reviewed in Jackson 1995; Belsham and Sonenberg 1996; Jackson et al. 1996). Viral IRES-dependent translation is efficient even when host cell cap-dependent translation is inhibited; this is mediated via the tertiary structure of the IRES, which is in turn stabilised by binding to cellular factors which interact directly with the translational machinery (Le and Maizel 1997; Gingras et al. 1999b; Hunt et al. 1999a, b; Hunt and Jackson 1999; Honda et al. 2000). IRES-dependent internal initiation is also facilitated by picornaviral-encoded proteases, which target eIF4GI and eIF4GII for cleavage (Lamphear et al. 1995; Gradi et al. 1998b; Svitkin et al. 1999), and consequently down regulate competing translation of cellular mRNA. In picornavirus-infected cells, the inhibition of translation correlates with the cleavage of eIF4GII and PABP, generating fragments favouring conditions for the IRES-dependent translation of viral mRNAs (Lloyd et al. 1987; Gradi et al. 1998b; Joachims et al. 1999; Kerekatte et al. 1999; Rhoads 1999; Svitkin et al. 1999). eIF4G is cleaved into two fragments, releasing the N-terminal fragment (which retains the binding to eIF4E and PABP) from the eIF4F complex (Fig. 2). Meanwhile, the C-terminal fragment of eIF4G, which retains the eIF4A binding site (Gingras et al. 1999b), mediates the eIF3 and RRM-dependent ribosome recruitment to the IRES (Pestova et al. 1996; Sizova et al. 1998) and functions efficiently in promoting internal initiation (Ohlmann et al. 1995, 1996, 1997; Borman et al. 1997; De Gregorio et al. 1998).

3.4.2 Potential IRES in Capped, Cellular mRNAs

The 5′ UTRs of a number of cellular mRNAs have features which are incompatible with efficient ribosome scanning (Kozak 1991). The 5′ UTRs of this class of mRNAs tend to be long and possess a high G + C content and the potential to form secondary structures; they also often show open reading frames (uORF) bounded by in-frame initiation and termination codons (reviewed in Kozak 1991; Van der Velden and Thomas 1999; Willis 1999). It is now clear that a number of these capped, cellular mRNAs have the ability to utilise internal ribosome binding to promote mRNA translation; however, unlike the case with viral IRES, cap-dependent and -independent translation of cellular mRNAs are not mutually exclusive. Recent DNA microarray technology has identified classes of cellular mRNAs which remain associated with polysomes following poliovirus infection when eIF4G integrity is compromised, and it has been estimated that as many as 3% of total cellular mRNAs may function in this manner (Johannes et al. 1999). Examples of capped mRNAs thought to possess IRES activity include those encoding c-myc (Stoneley et al. 1998, 2000a), FGF-2 (Vagner et al. 1995), BiP (Macejak and Sarnow 1991), VEGF (Akiri et al. 1998; Huez et al. 1998), the Kv1.4 cardiac voltage-gated potassium channel (Negulescu et al. 1998), IGF-II (Teerink et al. 1995), c-sis (Bernstein et al. 1995; Sella et al. 1999), DAP5 (Henis-Korenblit et al. 2000), ornithine decarboxylase (ODC; Gingras et al. 1999b), pim-1 (Johannes et al. 1999), connexin 43 (Schiavi et al. 1999), XIAP (Holcik et al. 1999), Apaf-1 (Coldwell et al. 2000), Gtx (Chappell et al. 2000), and the ubiquitous transcription factor, NRF (Oumard et al. 2000). Unlike their viral counterparts, there is little sequence homology between different cellular IRES (even a 9-nucleotide sequence in Gtx mRNA exhibits IRES activity which is enhanced when present in linked multiple copies; Chappell et al. 2000). They appear to function primarily at the 5′ end of the mRNA, although exceptions exist (e.g. the IRES of the β-subunit of mitochondrial H^+-ATP synthase which is located in the 3′ UTR of the mRNA; Izquierdo and Cuezva 2000).

Utilisation of the cellular IRES sequence on capped mRNAs may be regulated by a combination of alternative transcription from distinct promoters, translational control via IRES and uORF functions, and specific *trans*-acting factors (Teerink et al. 1995; Pyronnet et al. 1996; Shantz et al. 1996; Akiri et al. 1998; Carter et al. 1999; Pozner et al. 2000; Stoneley et al. 2000b). For example, the c-sis IRES is utilised preferentially during differentiation (Bernstein et al. 1995; Sella et al. 1999), the VEGF IRES under conditions of hypoxia (Stein et al. 1998), and the XIAP IRES under conditions of cell stress (Holcik et al. 1999). These are all conditions when eIF4F levels tend to be reduced in the cell, and are also conditions which promote "ribosomal subunit jumping" in adenovirus-infected cells (Yueh and Schneider 1996).

A number of important points on the role of cellular IRES sequences still await clarification, including the nature of the signal that promotes the switch from predominantly cap-dependent to cap-independent translation of cellular mRNAs. Moreover, it is far from clear how cellular IRES sequences actually function. Do they resemble viral IRES in interacting with canonical initiation factors involved in mRNA binding, e.g. eIF4G (Pestova et al. 1996; Kolupaeva et al. 1998), eIF4B (Meyer et al. 1995; Rust et al. 1999), or eIF3 (Sizova et al. 1998), or with other general or specific cellular proteins (Belsham and Sonenberg 1996; Roberts et al. 1998; Hunt and Jackson 1999; Hunt et al. 1999a; Stoneley et al. 2000b). An interesting possibility would be if the cellular IRES promotes translation in an analogous manner to that of transcriptional enhancers. Is there a role for leaky scanning (Kozak 1999) or "ribosomal subunit jumping" (Yueh and Schneider 1996) in cellular IRES function? Another question concerns the regulatory mechanisms involved, where the same mRNAs possess both cellular IRES sequences and upstream open reading frames that themselves appear to modulate translation, e.g. ODC, myc, pim-1. In particular, it would be interesting to determine whether IRES sequences and uORFs function in concert on the same mRNA to increase or decrease, respectively, its translation under different growth conditions.

Although cap-dependent and cap-independent translation are not mutually exclusive events on capped mRNA, it is rather surprising that increased expression of eIF4E or eIF4F complexes in the cell can also, in some cases, increase protein expression from IRES-containing mRNAs. Overexpression of eIF4E increased the synthesis of FGF-2 (Kevil et al. 1996; Nathan et al. 1997a), with IRES function apparently more efficient in transformed cells because more eIF4E was available to participate in translation initiation (Galy et al. 1999). Additionally, the inhibitory effects of the 5′ UTR on translation of the IRES-containing pim-1 (Hoover et al. 1997) and ODC mRNAs (Shantz et al. 1996) were relieved following eIF4E overexpression. ODC levels are also upregulated in activated T lymphocytes (Boal et al. 1993) at a time when eIF4F complexes are increased (Morley et al. 1993; Morley and Pain 1995b). This raises the broader question as to how overexpression of eIF4E in the cell really modulates gene expression. The original observations that overexpression of this factor resulted in deregulation of cell proliferation (De Benedetti and Rhoads 1990; Lazaris-Karatzas et al. 1990) led to the attractive hypothesis that enhanced eIF4F complex formation would selectively enhance translation of mRNAs encoding growth-regulatory proteins. Many of these mRNAs were known to have 5′ UTRs harbouring secondary structures that would impede cap-dependent initiation (Kozak 1991), and their translation would thus be expected to be highly dependent on the unwinding ability of the eIF4F complex. The hypothesis linking eIF4E levels to increased translation of such mRNAs was indeed strengthened by studies showing enhanced translation of structured reporter constructs (Koromilas et al. 1992) and selective increases in synthesis of ODC (Manzella et al. 1991) and cyclin D1 in eIF4E-overexpressing cells (Rosenwald et al. 1993). However, the effect on cyclin D1 synthesis was later found to be linked to increased nuclear export of the mRNA (Rousseau et al. 1996b), while the ability of increased levels of eIF4E to modulate translation of ODC mRNA is much more dependent on the presence of an upstream ORF than on the region of the 5′ UTR harbouring secondary structure (Shantz et al. 1996). Taken together with the puzzling observations discussed above on the relationships between changes in eIF4F and the utilisation of cellular IRES sequences, it seems likely that the the effects of eIF4E expression are much more complicated than originally thought.

4 Cell Death and the eIF4F Complex

4.1 Apoptosis: An Overview

Programmed cell death (apoptosis) is now recognised as an important physiological process involved in the regulation of cell and tissue growth, differentiation and developmental programmes. The molecular mechanisms of apoptosis have been the subject of intense research in recent years (reviewed in Cohen 1997; Nagata 1997; Peter et al. 1997; Rowan and Fisher 1997; Wyllie 1997; Jarpe et al. 1998; Kidd 1998; Porter 1999; Rathmell and Thompson 1999; Schulze-Osthoff et al. 1998; Stennicke and Salvesen 1999; Wallach et al. 1999; Wolf and Green 1999) and hence will only be covered superficially here. Briefly, cell death, which can be divided into a commitment phase and an execution phase, can be induced following the stimulation of specific cell surface receptors [such as the CD95 (Apo-1/Fas) antigen], or due to the lack of specific mitogenic growth factors, or following DNA damage. During the commitment phase of the apoptotic process, multi-protein complexes are assembled either at the membrane receptor [the death-inducing signalling complex (DISC)] or in the cytoplasm (aposomes; Cai et al. 1998). These complexes comprise various adapter proteins and one or more cysteine proteases (caspases); in each instance, a procaspase (usually caspase-8 or -9) is activated by proteolysis, leading to activation of further caspase cascades by limited proteolysis. During the execution phase, the downstream caspases (typically caspases-3 and -7), then target a variety of other cellular proteins for cleavage (Widmann et al. 1998), eventually resulting in the activation of DNA endonucleases, changes in cell permeability, cytoplasmic vacuolation and ultimately cell death.

Until recently, relatively little attention has been paid to the changes in protein synthesis, in terms of either overall translational activity or alterations in mRNA selection, that accompany the commitment and execution phases of apoptosis. However, new findings have begun to shed some light on this topic and provide insights into the possible mechanisms involved. This part of the chapter describes these developments and indicates the likely physiological significance of the data for our understanding of translational control during apoptosis. Collectively, these studies suggest that several proteins central to binding mRNA to the ribosome are direct or indirect targets for caspase-mediated regulation in vivo. However, it is not yet possible to rank the relative importance of these various events in mediating the inhibition of translation that occurs during apoptosis.

4.1.1
Apoptosis and Protein Synthesis Rates

The induction of apoptosis has been shown to be associated with a rapid, substantial, but incomplete, inhibition of protein synthesis in several cell types (reviewed in Clemens et al. 2000). Treatment of Jurkat T cells with anti-Fas antiserum results in a 60–70% decrease in the rate of protein synthesis within 2–4 h (Fig. 5A; Morley et al. 1998) and a loss of cell viability (Fig. 5B). The inhibition of protein synthesis is associated with a substantial decrease in the proportion of ribosomes in polysomes (Morley et al. 1998; Zhou et al. 1998), strongly suggesting that there is a block at the stage of polypeptide chain initiation.These events were prevented by the cell-permeable caspase inhibitor, z.VAD.FMK, indicating that caspase activity is involved. However, the mechanisms involved depend on the nature of the apoptotic inducer since the downregulation of translation caused by the DNA damaging agent etoposide was zVAD.FMK-insensitive (Morley et al. 1998).

4.2
Apoptosis and Initiation Factors Involved in Binding mRNA to the Ribosome

4.2.1
eIF4GI and eIF4GII

Recently, we (Clemens et al. 1998; Morley et al. 1998, 2000; Bushell et al. 1999, 2000a, b) and others (Marissen and Lloyd 1998) have demonstrated that a limited number of initiation factors are targets for specific degradation during apoptosis. With the human BJAB B cell lymphoma line, deprivation of serum growth factors or treatment with cycloheximide led to inhibition of protein synthesis and the progressive disappearance of eIF4GI and eIF4GII (Clemens et al. 1998; Bushell et al. 2000a). In contrast, under the same conditions, there were no major decreases in the levels of several other initiation factors, including eIF4E, eIF4A, PABP and 4E-BP1, and the phosphorylation of eIF4E decreased with time (Bushell et al. 2000a). However, there was a substantial increase in the association of 4E-BP1 with eIF4E at later times after induction of apoptosis in BJAB cells (Bushell et al. 2000a) and Jurkat cells (Morley et al. submitted).

Degradation of eIF4GI and eIF4GII occurred in a variety of cell types via both cell surface receptor-mediated and receptor-independent inducers of apoptosis (reviewed in Clemens et al. 2000). Treatment of Jurkat cells with anti-Fas antiserum promoted the inhibition of protein synthesis, temporally correlated with the disappearance of intact eIF4G and with the cleavage of the classical caspase substrate poly(ADP-ribose) polymerase, PARP (Fig. 5C). The cleavage of eIF4G was accompanied by the appearance of discrete breakdown products (Figs. 5C and 6A), termed FAGs (<u>F</u>ragment of <u>A</u>poptotic cleavage of

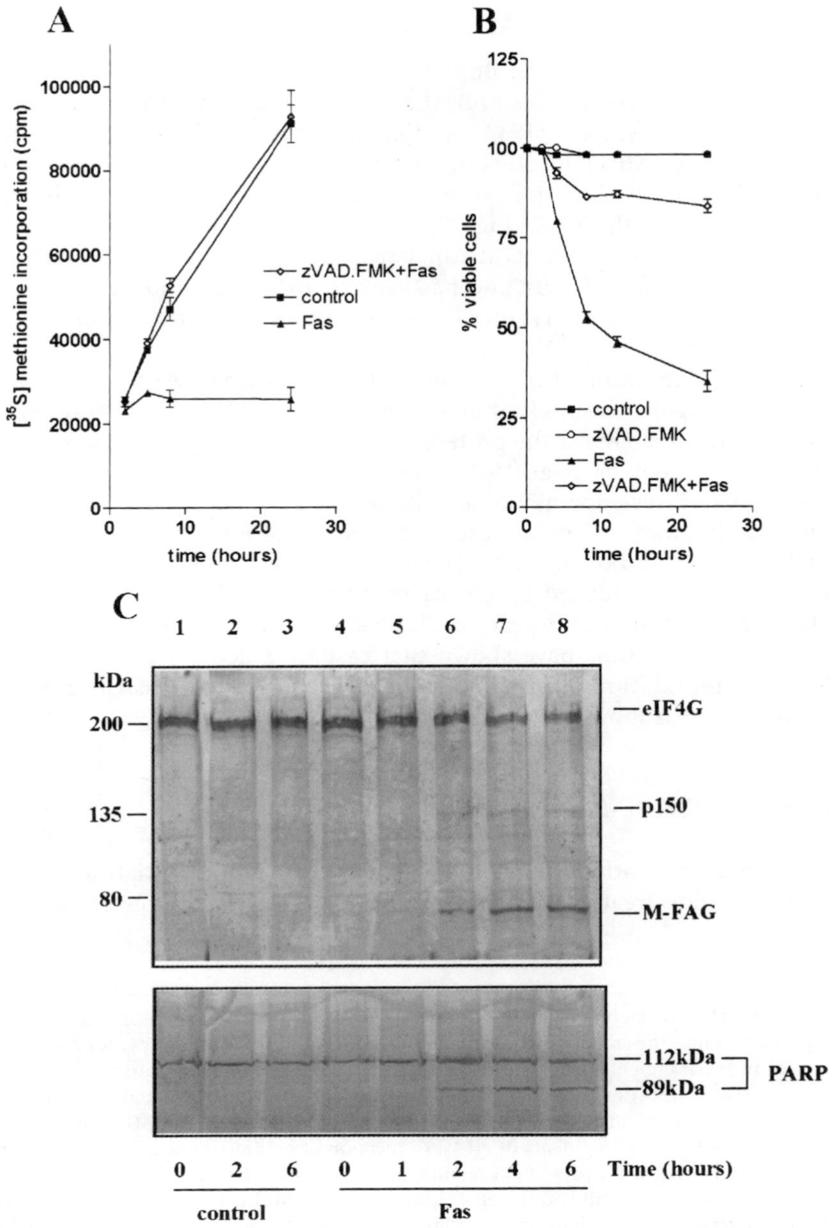

Fig. 5A–C. Apoptosis in Jurkat T cells is associated with the inhibition of protein synthesis and cleavage of eIF4G. **A** Jurkat T cells were preincubated in the absence or presence of 50 μM zVAD.FMK prior to incubation with or without 250 ng/ml anti-Fas antiserum. Rates of protein synthesis were estimated by pulse-labelling with [^{35}S] methionine for 30 min before harvesting at the times indicated. **B** Parallel cultures were used to determine cell viability, as estimated by trypan blue exclusion, and confirmed by FACS analysis (data not shown). **C** Jurkat cells were incubated in the absence or presence of 250 ng/ml anti-Fas antiserum for the times shown. Cell extracts were prepared and equal amounts of protein were resolved by SDS-PAGE; the cleavage of eIF4G (*upper panel*) and PARP (*lower panel*) were visualised by immunoblotting. (Morley 1997)

eIF4G; Clemens et al. 1998; Bushell et al. 2000a). N-FAG contains the PABP binding site, M-FAG retains both the eIF4E and eIF3/eIF4A interaction sites, and C-FAG, the second eIF4A binding site and the site responsible for the recruitment of Mnk1 to eIF4G (Fig. 6A). Using a battery of antisera, recent studies have shown that following cleavage of eIF4G, only M-FAG is retained in association with eIF4E (Fig. 6B; Bushell et al. 2000a). As M-FAG contains the eIF4E binding site, it is distinct from fragments of eIF4G generated in picornavirus-infected cells, and has been defined independently as the minimal sequence of eIF4G required to bind mRNA to ribosomes in an in vitro assay (Morino et al. 2000).

Both the inhibition of translation and the cleavage of eIF4GI was prevented by cell-permeable caspase inhibitors, with caspase-3 activity being both necessary and sufficient for the proteolysis of eIF4GI in vitro and in vivo (Morley et al. 1998; Clemens et al. 1998; Bushell et al. 1999). Although activation of the Fas/CD95 receptor in Jurkat cells resulted in increased activity of stress kinases, this was not required for the cleavage of eIF4GI and, rather surprisingly, eIF4E was not phosphorylated under these conditions. On the other hand, etoposide-induced apoptosis resulted in a robust phosphorylation of eIF4E in the same cells (Morley et al. 1998). Additionally, studies with caspase-8 mutant Jurkat cells have shown that caspase-8 activity is required for the long-term inhibition of protein synthesis in response to Fas/CD95 activation (Morley et al., submitted).

4.2.2
eIF4B, eIF3p35, 4E-BP1 and eIF2α

Further investigations into the integrity of other initiation factors in cycloheximide-treated cells undergoing apoptosis have now shown that, in

Fig. 6A,B. M-FAG retains the ability to interact with eIF4E during apoptosis. **A** Diagrammatic representation of the caspase-3 cleavage sites in eIF4GI and the binding sites for other initiation factors in the fragments generated. Note that the L protease of foot-and-mouth-disease virus bisects eIF4GI, whilst caspase-3 cleaves at two sites (after residues 492 and 1136, Bushell et al. 2000a; numbering of Imataka et al. 1998) to produce three fragments, N-FAG, M-FAG and C-FAG. Also indicated are the fragments of eIF4G or location of peptide sequences used to raise antisera used in **B**. The M-FAG fragment retains the ability to bind eIF4E, eIF3 and eIF4A but loses the binding sites for PABP and the eIF4E protein kinase, Mnk1. **B** Exponentially growing BJAB cells were treated with cycloheximide (100 μg/ml) for the times indicated and cell extracts were prepared. Equal quantities of protein from each extract were subjected to SDS gel electrophoresis, followed by immunoblotting for eIF4G (with antiserum W, RL, or E, as indicated) and eIF4E (*left panel*). Molecular mass markers and the locations of the N-terminal fragment (N-FAG), the middle fragment (M-FAG), and the C-terminal (C-FAG) fragment of eIF4GI are indicated. Aliquots of cell extract containing equal amounts of protein prepared as above were subjected to affinity chromatography on m^7GTP-Sepharose. eIF4E and associated proteins were recovered and resolved by SDS-PAGE and immunoblotting for eIF4e and eIF4G and its cleavage fragments (using antisera W, RL and E), as indicated (*right panel*). (Data from Bushell et al. 2000a)

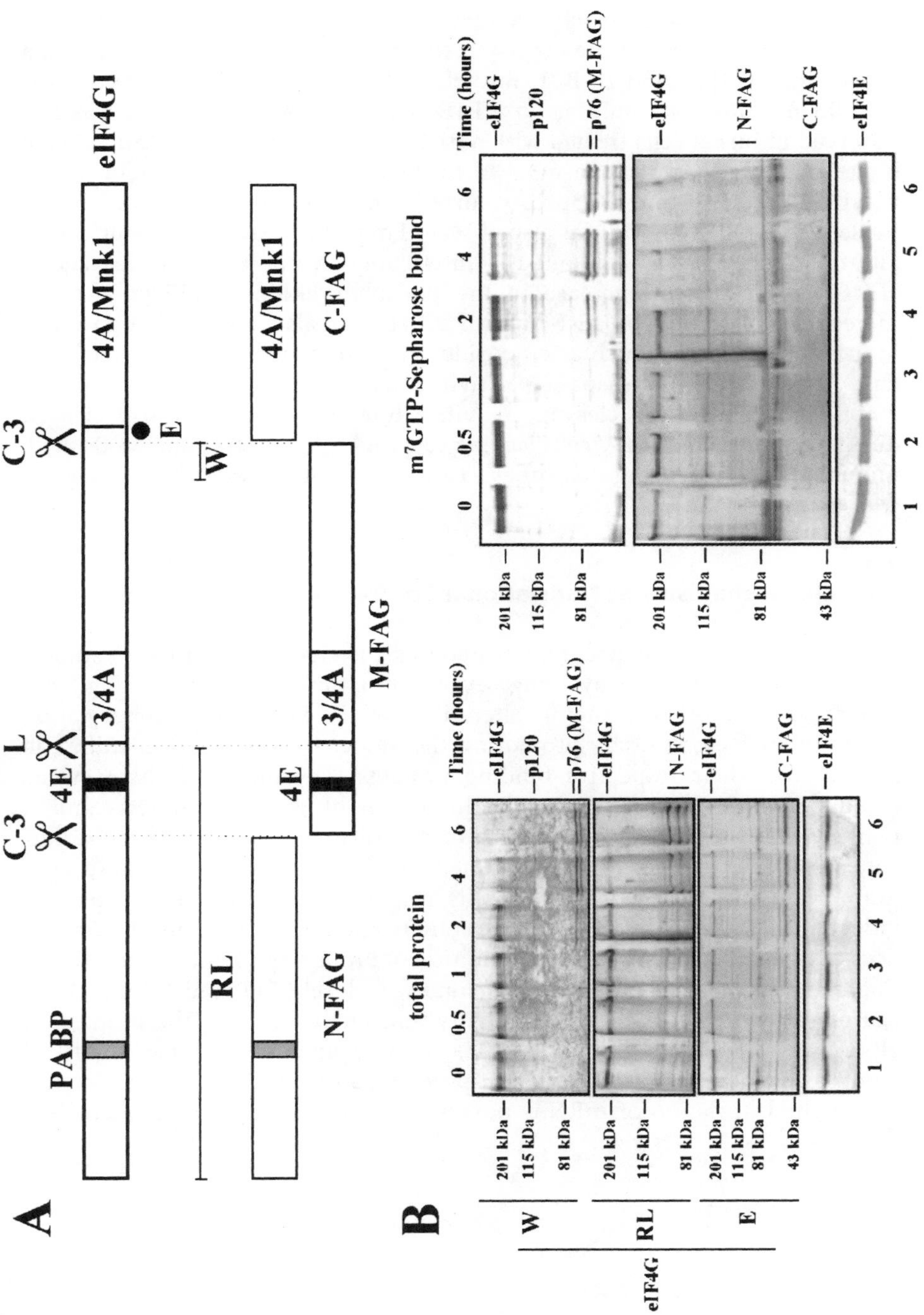
A
PABP
C-3
4E
L
3/4A
C-3
E
4A/Mnk1
eIF4G1
RL
W
N-FAG
4E
3/4A
M-FAG
4A/Mnk1
C-FAG
B
total protein
m7GTP-Sepharose bound
Time (hours)
eIF4G
p120
p76 (M-FAG)
N-FAG
C-FAG
eIF4E
201 kDa
115 kDa
81 kDa
43 kDa
W
RL
E
eIF4G

addition to eIF4GI and eIF4GII, there was selective cleavage of eIF4B, the p35 subunit of eIF3 (none of the other ten subunits of eIF3 were affected), and a population of eIF2α and 4E-BP1 (Bushell et al. 2000a, b). In the last case, cleavage did not affect its binding to eIF4E (Bushell et al. 2000a) and was not observed in Jurkat cells treated with anti-Fas antiserum or in the reticulocyte lysate incubated with caspase-3 (M. Bushell and S. Morley, unpubl. data). Cleavage of eIF4B and eIF3p35 occurred with delayed kinetics relative to those seen for eIF4GI, and was also observed in response to serum deprivation and treatment of cells with anti-Fas antiserum or etoposide. The cleavage of eIF4GI, the transient increase in the phosphorylation of eIF2α, and the increased association of 4E-BP1 with eIF4E was also observed with other inducers of apoptosis, such as etoposide or staurosporine (S. Morley, unpubl. data). As with eIF4GI, caspase-3 activity was both necessary and sufficient for eIF3p35 and eIF4B cleavage in vitro; however eIF4B was also cleaved during apoptosis in MCF-7 cells lacking caspase-3 (I.W. Jeffrey and M. Bushell, unpubl. data), suggesting that other caspases may also target this protein for degradation.

4.3
Possible Mechanisms of Translational Control During Apoptosis

As indicated above, the progress of apoptosis is characterized by a complex programme of changes involving several initiation factors, including: specific fragmentation of proteins; alterations in the state of phosphorylation of initiation factors; or by promoting the association of selected initiation factors with their respective binding proteins. Any or all of these events could potentially contribute to the inhibition of protein synthesis and it is likely that the relative importance of the various changes may be different at distinct stages of development of the apoptotic response. For example, in anti-Fas treated Jurkat cells, it is been shown that the caspase-8-independent increase in eIF2α kinase activity and eIF2α phosphorylation suffices to cause a transient, general inhibition of protein synthesis at early times. The consequent disassembly of polysomes is closely followed by the cleavage of eIF4G, the loss of $p70^{S6K}$ activity and an increase in the binding of 4E-BP1 to eIF4E. Inhibition of $p70^{S6K}$ activity and a decreased ability of eIF4E to participate in the initiation process add to this effect but may also favour the translation of mRNAs possessing a functional IRES element (see below).

4.3.1
Cleavage of eIF2α, eIF4B, eIF3p35 and Increased Binding of 4E-BP1 to eIF4E

One of the earliest effects of apoptotic inducers on cells is an increase in the phosphorylation of eIF2α. In anti-Fas-treated Jurkat cells, this effect is transient but correlates with the onset of the inhibition of protein synthesis. In these and other cell types, as well as becoming more highly phosphorylated, a population of eIF2α is also cleaved to give rise to a C-terminally truncated form (Satoh et al. 1999; Bushell et al. 2000b; Marissen et al. 2000). Although it can still be phosphorylated by the protein kinase activated by double-stranded RNA (PKR), the truncated form of eIF2α has been reported to block the PKR-mediated suppression of reporter gene expression (Satoh et al. 1999), possibly by allowing eIF2B-independent GDP exchange on the truncated eIF2 complex (Marissen et al. 2000). Thus the caspase-mediated cleavage of eIF2α may provide a mechanism for reversing the effects of phosphorylation on the function of initiation factor eIF2 during apoptosis. Alternative possibilities which need to be addressed are that eIF2α phosphorylation may actually promote the progression of cell death by facilitating the translation of the pro-apoptotic proteins such as Bax and Fas (Cohen 1997; Wyllie 1997; Porter 1999), or by blocking the synthesis of anti-apoptotic gene products (Marissen et al. 2000). As Bax and Fas mRNAs possess uORFs, and increased synthesis of Bax protein coincides with increased levels of eIF2α phosphorylation, it is possible that translational control of Bax mRNA occurs during apoptosis in a manner similar to that described for GCN4 in *S. cerevisiae* (Hinnebusch 1997; Jagus et al. 1999). Further work is required to resolve these possibilities.

The fragment of eIF4B which is generated in apoptotic cells retains the domain required for self-association and for interaction with eIF3 (Methot et al. 1996). Although the cleaved form of eIF4B can still interact with the cap binding complex (Bushell et al. 2000b), it is too early to say whether the cleavage of eIF4B has a functional significance. Similarly, although the role of the p35 subunit of eIF3 in the function of this multi-subunit factor remains unknown (Block et al. 1994), the possibility that eIF3p35 cleavage contributes to the downregulation or reprogramming of translation cannot be ruled out.

In Jurkat cells, the CD95-mediated inhibition of signalling via the PKB/Akt pathway has been associated with decreased phosphorylation and cleavage of $p70^{S6K}$ and increased binding of 4E-BP1 to eIF4E (Morley et al., submitted). Downregulation of $p70^{S6K}$ would be predicted to specifically result in a decrease in the translation of mRNAs containing a 5′ polypyrimidine tract (Meyuhas et al. 1996), while the dephosphorylation of 4E-BP1 is predicted to result in the sequestration of eIF4E away from eIF4G, potentially resulting in a general inhibition of translation.

4.3.2
Cleavage of eIF4G

Cleavage of eIF4GI and eIF4GII is a relatively early event in apoptosis, resulting in the production of a modified form of eIF4F, containing eIF4E and eIF4A, but with the central M-FAG fragment in place of full length eIF4GI. This complex, which is distinct from that observed during cell growth (Fig. 7) remains stable for several hours in apoptosing cells (Fig. 5C) before M-FAG is further degraded with the loss of the eIF4E binding site. The decrease in the phosphorylation state of eIF4E during apoptosis may also be a consequence of eIF4G cleavage because the binding site for the eIF4E kinase, Mnk1, is not present in M-FAG (Pyronnet et al. 1999).

In an analogous manner to that described for poliovirus infection of cells, the accumulation of the caspase-generated eIF4G cleavage fragments during apoptosis may directly influence mRNA translation rates in vivo under conditions where the interaction between the 5′ and 3′ ends of the mRNA has been disrupted (reviewed in Clemens et al. 2000). Unlike the situation in picornavirus-infected cells, where cleavage of eIF4G separates the site that interacts with eIF4E from other functional regions, this modified eIF4F complex may still be able to support either cap-dependent (Morino et al. 2000) or -independent initiation (De Gregorio et al. 1998). Additionally, it is also possible that one or more of the eIF4G cleavage fragments may show a gain of function. For example N-FAG, which can be stable in apoptosing cells for at least 24 h (Marissen and Lloyd 1998), may impair poly(A)-dependent initiation or its regulation by sequestering PABP (Imataka et al. 1998), and C-FAG may interfere with eIF4E phosphorylation by titrating out Mnk1 (Pyronnet et al. 1999).

Several cellular mRNAs demonstrated to be capable of translation by a cap-independent mechanism encode proteins intimately associated with cell death. These include the inhibitor of apoptosis XIAP (Holcik et al. 1999), c-myc (Stoneley et al. 1998, 2000a), Apaf-1 (Coldwell et al. 2000) and DAP5 (Henis-Korenblit et al. 2000). Furthermore, the translation of these mRNAs is maintained during apoptosis under conditions when it is predicted that the modified form of the eIF4F complex would be present and general protein synthesis rates are decreased. Indeed, the caspase-mediated cleavage product of DAP5 can apparently enhance its own IRES-mediated translation during apoptosis (Henis-Korenblit et al. 2000). It is predicted that the maintained expression of proteins of the IAP family would limit the apoptotic response (Deveraux et al. 1999; Deveraux and Reed 1999), whereas synthesis of DAP5 or Apaf-1 would promote apoptosis by providing a positive feedback loop. Clearly, further work is required to address the possible roles of the eIF4G fragments N-FAG, M-FAG or C-FAG in mediating cap-dependent and/or -independent translation in apoptosing cells and to determine whether the continued translation of specific mRNAs such as those described above is influenced by these proteins.

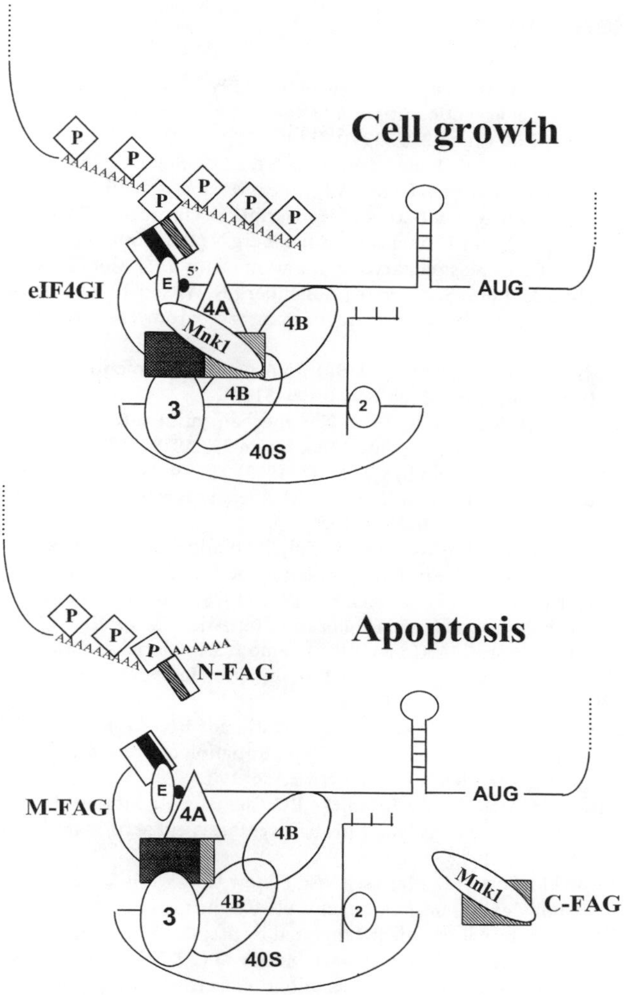

Fig. 7. A comparison of eIF4F complexes assembled on capped mRNAs during cell growth and apoptosis. Interaction between individual initiation factors and the 5′ and 3′ ends of the mRNA during cell growth or cell death (see text and Fig. 2 for details)

Acknowledgements. I am grateful to Jenny Pain and Linda McKendrick for critical reading of this manuscript and to all members of my lab, past and present for helpful discussions. The research in my lab is supported by generous project and equipment grants from The Wellcome Trust (040800, 050703, 056778, 047029, 054256, 057494, 058915). In addition, the author is a Senior Research Fellow of The Wellcome Trust.

References

Abid MR, Li Y, Anthony C, De Benedetti A (1999) Translational regulation of ribonucleotide reductase by eukaryotic initiation factor 4E links protein synthesis to the control of DNA replication. J Biol Chem 274:35991–35998

Akiri G, Nahari D, Finkelstein Y, Le SY, Elroy-Stein O, Levi BZ (1998) Regulation of vascular endothelial growth factor (VEGF) expression is mediated by internal initiation of translation and alternative initiation of transcription. Oncogene 17:227–236

Altmann M, Edery I, Trachsel H, Sonenberg N (1988) Site-directed mutagenesis of the tryptophan residues in yeast eukaryotic initiation factor 4E. J Biol Chem 263:17229–17232

Altmann M, Blum S, Pelletier J, Sonenberg N, Wilson TMA, Trachsel H (1990) Translation initiation factor-dependent extracts from Saccharomyces cerevisiae. Biochim Biophys Acta 1050: 155–159

Belsham GJ, Sonenberg N (1996) RNA-protein interactions in regulation of picornavirus RNA translation. Microbiol Rev 60:499–511

Benne R, Hershey JWB (1978) The mechanism of action of protein synthesis initiation factors from rabbit reticulocytes. J Biol Chem 253:3078–3087

Bernstein J, Shefler I, Elroy-Stein O (1995) The translational repression mediated by the platelet-derived growth factor 2/c-sis mRNA leader is relieved during megakaryocytic differentiation. J Biol Chem 270:10559–10565

Bi X, Goss DJ (2000) Wheat germ poly(A) binding protein increases the ATPase and the RNA helicase activity of translation initiation factors eIF4A, eIF4B and eIF-iso4F. J Biol Chem (in press)

Block KL, Vornlocher HP, Hershey JWB (1998) Characterization of cDNAs encoding the p44 and p35 subunits of human translation initiation factor eIF3. J Biol Chem 273:31901–31908

Boal TR, Chiorini JA, Cohen RB, Miyamoto S, Frederickson RM, Sonenberg N, Safer B (1993) Regulation of eukaryotic translation initiation factor expression during T-cell activation. Biochim Biophys Acta 1176:257–264

Borman AM, Kirchweger R, Ziegler E, Rhoads RE, Skern T, Kean KM (1997) eIF4G and its proteolytic cleavage products: Effect on initiation of protein synthesis from capped, uncapped, and IRES-containing mRNAs. RNA 3:186–196

Bushell M, McKendrick L, Janicke RU, Clemens MJ, Morley SJ (1999) Caspase-3 is necessary and sufficient for cleavage of protein synthesis eukaryotic initiation factor 4G during apoptosis. FEBS Lett 451:332–336

Bushell M, Poncet D, Marissen WE, Flotow H, Lloyd RE, Clemens MJ, Morley SJ (2000a) Cleavage of polypeptide initiation factor eIF4GI during apoptosis in lymphoma cells: characterisation of an internal fragment generated by caspase-3 cleavage. Cell Death Differ (in press)

Bushell M, Wood W, Clemens MJ, Morley SJ (2000b) Changes in the integrity and association of eukaryotic protein synthesis initiation factors during apoptosis. Eur J Biochem 267:1083–1091

Cai JY, Yang J, Jones DP (1998) Mitochondrial control of apoptosis: the role of cytochrome c. Biochim Biophys Acta 1366:139–149

Carter PS, Jarquin-Pardo M, De Benedetti A (1999) Differential expression of Myc1 and Myc2 isoforms in cells transformed by eIF4E: evidence for internal ribosome repositioning in the human c-myc 5′ UTR. Oncogene 18:4326–4335

Chappell SA, Edelman GM, Mauro VP (2000) A 9-nt segment of cellular mRNA can function as an internal ribosome entry site (IRES) and when present in linked multiple copies greatly enhances IRES activity. Proc Natl Acad Sci USA 97:1536–1541

Clemens MJ, Bushell M, Morley SJ (1998) Degradation of eukaryotic polypeptide chain initiation factor (eIF) 4G in response to induction of apoptosis in human lymphoma cell lines. Oncogene 17:2921–2931

Clemens MJ, Bushell M, Jeffrey IW, Pain VM, Morley SJ (2000) Translation initiation factor modifications and the regulation of protein synthesis in apoptotic cells. Cell Death Differ (in press)

Cohen GM (1997) Caspases: the executioners of apoptosis. Biochem J 326:1–16

Coldwell M, Mitchell SA, Stoneley M, MacFarlane M, Willis AE (2000) Initiation of Apaf-1 translation by internal ribosome entry. Oncogene 19:899–905

Craig AWB, Haghighat A, Yu ATK, Sonenberg N (1998) Interaction of polyadenylate-binding protein with the eIF4G homologue PAIP enhances translation. Nature 392:520–523

De Benedetti A, Rhoads RE (1990) Overexpression of eukaryotic protein synthesis initiation factor 4E in HeLa cells results in aberrant growth and morphology. Proc Natl Acad Sci USA 87:8212–8216

De Gregorio E, Preiss T, Hentze MW (1998) Translational activation of uncapped mRNAs by the central part of human eIF4G is 5′ end-dependent. RNA 4:828–836

De la Cruz J, Kressler D, Linder P (1999) Unwinding RNA in Saccharomyces cerevisiae: DEAD-box proteins and related families. Trends Biochem Sci 24:192–198

Denton RM, Tavaré JM (1995) Does mitogen-activated-protein kinase have a role in insulin action? The cases for and against. Eur J Biochem 227:597–611

Dever TE (1999) Translation initiation: adept at adapting. Trends Biochem Sci 24:398–403

Deveraux QL, Reed TC (1999) IAP family proteins – suppressors of apoptosis. Genes Dev 13: 239–252

Deveraux QL, Leo E, Stennicke HR, Welsh K, Salvesen GS, Reed JC (1999) Cleavage of human inhibitor of apoptosis protein XIAP results in fragments with distinct specificities for caspases. EMBO J 18:5242–5251

Diggle TA, Moule SK, Avison MB, Flynn A, Foulstone EJ, Proud CG, Denton RM (1996) Both rapamycin-sensitive and -insensitive pathways are involved in the phosphorylation of the initiation factor-4E-binding protein (4E-BP1) in response to insulin in rat epididymal fat-cells. Biochem J 316:447–453

Duncan RF (1996) Translational control during heat shock. In: Hershey JWB, Mathews MB, Sonenberg N (eds) Translational control. Cold Spring Harbor Laboratory Press, Plainsview, NY, pp 271–294

Edery IM, Humbelin M, Darveau A, Lee KA, Milburn S, Hershey JW, Trachsel H (1983) Involvement of eukaryotic initiation factor 4A in the cap recognition process. J Biol Chem 258:11398–11403

Flynn A, Proud CG (1995) Serine 209, not serine 53, is the major site of phosphorylation in initiation factor eIF-4E in serum-treated Chinese hamster ovary cells. J Biol Chem 270: 21684–21688

Flynn A, Proud CG (1996a) The role of eIF4 in cell proliferation. Cancer Surv 27:293–310

Flynn A, Proud CG (1996b) Insulin-stimulated phosphorylation of initiation factor 4E is mediated by the MAP kinase pathway. FEBS Lett 389:162–166

Fraser CS, Pain VM, Morley SJ (1999a) The association of initiation factor 4F with poly(A)-binding protein is enhanced in serum-stimulated *Xenopus* kidney cells. J Biol Chem 274: 196–204

Fraser CS, Pain VM, Morley SJ (1999b) Cellular stress in *Xenopus* kidney cells enhances the phosphorylation of eukaryotic translation initiation factor (elF)4E and the association of elF4F with poly(A)-binding protein. Biochem J 342:519–526

Gallie DR (1996) Translational control of cellular and viral mRNAs. Plant Mol Biol 32:145–158

Gallie DR, Traugh JA (1994) Serum and insulin regulate cap function in 3T3-L1 cells. J Biol Chem 269:7174–7179

Galy B, Maret A, Prats AC, Prats H (1999) Cell transformation results in the loss of the density-dependent translational regulation of the expression of fibroblast growth factor 2 isoforms. Cancer Res 59:165–171

Gautsch TA, Anthony JC, Kimball SR, Paul GL, Layman DK, Jefferson LS (1998) Availability of eIF4E regulates skeletal muscle protein synthesis during recovery from exercise. Am J Physiol Cell Physiol 274:C406–C414

Gingras A-C, Gygi SP, Raught B, Polakiewicz RD, Abraham RT, Hoekstra MF, Aebersold R, Sonenberg N (1999a) Regulation of 4E-BP1 phosphorylation: a novel two-step mechanism. Genes Dev 13:1422–1437

Gingras A-C, Raught B, Sonenberg N (1999b) eIF4 initiation factors: effectors on mRNA recruitment to ribosomes and regulators of translation. Annu Rev Biochem 68:913–963

Goyer C, Altmann M, Lee HS, Blanc A, Deshmukh M, Woolford JL Jr, Trachsel H, Sonenberg N (1993) TIF4631 and TIF4632: Two yeast genes encoding the high-molecular-weight subunits of the cap-binding protein complex (eukaryotic initiation factor 4F) contain an RNA recognition motif-like sequence and carry out an essential function. Mol Cell Biol 13:4860–4874

Gradi A, Imataka H, Svitkin YV, Rom E, Raught B, Morino S, Sonenberg N (1998a) A novel functional human eukaryotic translation initiation factor 4G. Mol Cell Biol 18:334–342

Gradi A, Svitkin YV, Imataka H, Sonenberg N (1998b) Proteolysis of human eukaryotic translation initiation factor eIF4GII, but not eIF4GI, coincides with the shutoff of host protein synthesis after poliovirus infection. Proc Natl Acad Sci USA 95:11089–11094

Gray NK, Wickens M (1998) Control of translation initiation in animals. Annu Rev Cell Dev Biol 14:399–458

Grifo JA, Abramson RD, Sutter CA, Merrick WC (1984) RNA-stimulated ATPase activity of eukaryotic initiation factors. J Biol Chem 259:8648–8654

Grifo JA, Tahara SM, Morgan MA, Shatkin AJ, Merrick WC (1983) New initiation factor activity required for globin mRNA translation. J Biol Chem 258:5804–5810

Haghighat A, Sonenberg N (1997) eIF4G dramatically enhances the binding of eIF4E to the mRNA 5′-cap structure. J Biol Chem 272:21677–21680

Haghighat A, Mader S, Pause A, Sonenberg N (1995) Repression of cap-dependent translation by 4E-binding protein 1: Competition with p220 for binding to eukaryotic initiation factor-4E. EMBO J 14:5701–5709

Heesom KJ, Denton RM (1999) Dissociation of the eukaryotic initiation factor-4E/4E-BP1 complex involves phosphorylation of 4E-BP1 by an mTOR-associated kinase. FEBS Lett 457:489–493

Heesom KJ, Avison MB, Diggle TA, Denton RM (1998) Insulin-stimulated kinase from rat fat cells that phosphorylates initiation factor 4E-binding protein 1 on the rapamycin-insensitive site (serine-111). Biochem J 336:39–48

Henis-Korenblit S, Levy-Strumpf N, Goldstaub D, Kimchi A (2000) A novel form of DAP5 protein accumulates in apoptotic cells as a result of caspase cleavage and internal ribosome entry site-mediated translation. Mol Cell Biol 20:496–506

Hinnebusch AG (1997) Translational regulation of yeast GCN4 – a window on factors that control initiator-tRNA binding to the ribosome. J Biol Chem 272:21661–21664

Holcik M, Lefebvre C, Yeh C, Chow T, Korneluk RG (1999) A new internal-ribosome-entry-site motif potentiates XIAP-mediated cytoprotection. Nat Cell Biol 1:190–192

Honda M, Kaneko S, Matsushita E, Kobayashi K, Abell G, Lemon SM (2000) Cell cycle regulation of hepatitis C virus internal ribosome entry site-directed translation. Gastroenterology 118:152–162

Hoover DS, Wingett DG, Zhang J, Reeves R, Magnuson NS (1997) Pim-1 protein expression is regulated by its 5′-untranslated region and translation initiation factor eIF-4E. Cell Growth Differ 8:1371–1380

Huez I, Créancier L, Audigier S, Gensac MC, Prats AC, Prats H (1998) Two independent internal ribosome entry sites are involved in translation initiation of vascular endothelial growth factor mRNA. Mol Cell Biol 18:6178–6190

Hunt SL, Jackson RJ (1999) Polypyrimidine-tract binding protein (PTB) is necessary, but not sufficient, for efficient internal initiation of translation of human rhinovirus-2 RNA. RNA 5:344–359

Hunt SL, Hsuan JJ, Totty N, Jackson RJ (1999a) unr, a cellular cytoplasmic RNA-binding protein with five cold-shock domains, is required for internal initiation of translation of human rhinovirus RNA. Genes Dev 13:437–448

Hunt SL, Skern T, Liebig HD, Kuechler E, Jackson RJ (1999b) Rhinovirus 2A proteinase mediated stimulation of rhinovirus RNA translation is additive to the stimulation effected by cellular RNA binding proteins. Virus Res 62:119–128

Imataka H, Sonenberg N (1997) Human eukaryotic translation initiation factor 4G (eIF4G) possesses two separate and independent binding sites for eIF4A. Mol Cell Biol 17:6940–6947

Imataka H, Gradi A, Sonenberg N (1998) A newly identified N-terminal amino acid sequence of human eIF4G binds poly(A)-binding protein and functions in poly(A)-dependent translation. EMBO J 17:7480–7489

Izquierdo JM, Cuezva JM (2000) Internal ribosome entry site functional activity of the 3′-untranslated region of the mRNA for the beta subunit of mitochondrial H+ATP synthase. Biochem J 346:849–855

Jackson RJ (1995) A comparative overview of initiation site selection mechanisms. In: Hershey JWB, Mathews MB, Sonenberg N (eds) Translational control. Cold Spring Harbor Laboratory Press, Plainsview, NY, pp 71–112

Jackson RJ, Wickens M (1997) Translational controls impinging on the 5′-untranslated region and initiation factor proteins. Curr Opin Genet Dev 7:233–241

Jackson RJ, Hunt SL, Reynolds JE, Kaminski A (1995) Cap-dependent and cap-independent translation: operational distinctions and mechanistic interpretations. In: Sarnow P (ed) Cap-independent translation. Current topics in microbiology and immunology. Springer, Berlin Heidelberg New York, pp 251–290

Jacobson A (1996) Poly(A) metabolism and translation: the closed loop model. In: Hershey JWB, Mathews MB, Sonenberg N (eds) Translational control. Cold Spring Harbor Laboratory Press, Plainsview, NY, pp 451–480

Jagus R, Joshi B, Barber GN (1999) PKR, apoptosis and cancer. Int J Biochem Cell Biol 31:123–138

Jarpe MB, Widmann C, Knall C, Schlesinger TK, Gibson S, Yujiri T, Fanger GR, Gelfand EW, Johnson GL (1998) Anti-apoptotic versus pro-apoptotic signal transduction: checkpoints and stop signs along the road to death. Oncogene 17:1475–1482

Joachims M, Van Breugel PC, Lloyd RE (1999) Cleavage of poly(A)-binding protein by enterovirus proteases concurrent with inhibition of translation in vitro. J Virol 73:718–727

Johannes G, Carter MS, Eisen MB, Brown PO, Sarnow P (1999) Identification of eukaryotic mRNAs that are translated at reduced cap binding complex eIF4F concentrations using a cDNA microarray. Proc Natl Acad Sci USA 96:13118–13123

Joshi B, Cai A-L, Keiper BD, Minich WB, Mendez R, Beach CM, Stepinski J, Stolarski R, Darzynkiewicz E, Rhoads RE (1995) Phosphorylation of eukaryotic protein synthesis initiation factor 4E at Ser-209. J Biol Chem 270:14597–14603

Joshi-Barve S, Rychlik W, Rhoads RE (1990) Alteration of the major phosphorylation site of eukaryotic protein synthesis initiation factor 4E prevents its association with the 48 S initiation complex. J Biol Chem 265:2979–2983

Kerekatte V, Keiper BD, Badorff C, Cai AL, Knowlton KU, Rhoads RE (1999) Cleavage of poly(A)-binding protein by Coxsackie virus 2 A protease in vitro and in vivo: Another mechanism for host protein synthesis shutoff. J Virol 73:709–717

Kessler SH, Sachs AB (1998) RNA recognition motif 2 of yeast Pab1p is required for its functional interaction with eukaryotic translation initiation factor 4G. Mol Cell Biol 18:51–57

Kevil CG, De Benedetti A, Payne DK, Coe LL, Laroux FS, Alexander JS (1996) Translational regulation of vascular permeability factor by eukaryotic initiation factor 4E: Implications for tumor angiogenesis. Int J Cancer 65:785–790

Kidd VJ (1998) Proteolytic activities that mediate apoptosis. Annu Rev Physiol 60:533–573

Kimball SR, Horetsky RL, Jefferson LS (1998) Implication of eIF2B rather than eIF4E in the regulation of global protein synthesis by amino acids in L6 myoblasts. J Biol Chem 273: 30945–30953

Kimball SR, Jurasinski CV, Lawrence JC, Jefferson LS (1997) Insulin stimulates protein synthesis in skeletal muscle by enhancing the association of eIF-4E and eIF-4G. Am.J.Physiol.Cell Physiol. 272:C754–C759

Kirchweger R, Ziegler E, Lamphear BJ, Waters D, Liebig H-D, Sommergruber W, Sobrino F, Hohenadl C, Blaas D, Rhoads RE, Skern T (1994) Foot-and-mouth disease virus leader proteinase: Purification of the Lb form and determination of its cleavage site on eIF-4 gamma. J Virol 68:5677–5684

Kleijn M, Scheper GC, Voorma HO, Thomas AAM (1998) Regulation of translation initiation factors by signal transduction. Eur J Biochem 253:531–544

Kolupaeva VG, Pestova TV, Hellen CUT, Shatsky IN (1998) Translation eukaryotic initiation factor 4G recognises a specific structural element within the internal ribosome entry site of encephalomyocarditis virus RNA. J Biol Chem 273:18599–18604

Koromilas AE, Lazaris-Karatzas A, Sonenberg N (1992) mRNAs containing extensive secondary structure in their 5′ non-coding region translate efficiently in cells overexpressing initiation factor eIF-4E. EMBO J 11:4153–4158

Kozak M (1991) An analysis of vertebrate mRNA sequences: Intimations of translational control. J Cell Biol 115:887–903

Kozak M (1999) Initiation of translation in prokaryotes and eukaryotes. Gene 234:187–208

Lamphear BJ, Panniers R (1990) Cap binding protein complex that restores protein synthesis in heat-shocked Ehrlich cell lysates contains highly phosphorylated eIF-4E. J Biol Chem 265:5333–5336

Lamphear BJ, Panniers R (1991) Heat shock impairs the interaction of cap-binding protein complex with 5′ mRNA cap. J Biol Chem 266:2789–2794

Lamphear BJ, Kirchweger R, Skern T, Rhoads RE (1995) Mapping of functional domains in eukaryotic protein synthesis initiation factor 4G (eIF4G) with picornaviral proteases -Implications for cap-dependent and cap-independent translational initiation. J Biol Chem 270:21975–21983

Lang CH, Wu DQ, Frost RA, Jefferson LS, Vary TC, Kimball SR (1999) Chronic alcohol feeding impairs hepatic translation initiation by modulating eIF2 and eIF4E. Am J Physiol Endocrinol Metab 277:E805–E814

Laroia G, Cuesta R, Brewer G, Schneider RJ (1999) Control of mRNA decay by heat-shock-ubiquitin-proteosome pathway. Science 284:499–502

Lawrence JC Jr, Abraham RT (1997) PHAS/4E-BPs as regulators of mRNA translation and cell proliferation. Trends Biochem Sci 22:345–349

Lazaris-Karatzas A, Montine KS, Sonenberg N (1990) Malignant transformation by a eukaryotic initiation factor subunit that binds to mRNA 5′ cap. Nature 345:544–547

Le H, Tanguay RL, Balasta ML, Wei CC, Browning KS, Metz AM, Goss DJ, Gallie DR (1997) Translation initiation factors eIF-iso4G and eIF-4B interact with the poly(A)-binding protein and increase its RNA binding activity. J Biol Chem 272:16247–16255

Le SY, Maizel JV (1998) Evolution of a common structural core in the internal ribosome entry sites of picornavirus. Virus Genes 16:25–38

Li BDL, Liu L, Dawson M, De Benedetti A (1997) Overexpression of eukaryotic initiation factor 4E (eIF4E) in breast carcinoma. Cancer 79:2385–2390

Lin T-A, Kong X, Haystead TAJ, Pause A, Belsham G, Sonenberg N, Lawrence JC Jr (1994) PHAS-I as a link between mitogen-activated protein kinase and translation initiation. Science 266:653–656

Linder P, Slonimski PP (1989) An essential yeast protein, encoded by duplicated genes TIF1 and TIF2 and homologous to the mammalian translation initiation factor eIF-4A, can suppress a mitochondrial missense mutation. Proc Natl Acad Sci USA 86:2286–2290

Linder P, Lasko PF, Ashburner M, Leroy P, Neilsen PJ, Nishi K, Schnier J, Slonimski PP (1989) Birth of the DEAD box. Nature 337:121–122

Lloyd RE, Jense HG, Ehrenfeld E (1987) Restriction of translation of capped mRNA in vitro as a model for poliovirus-induced inhibition of host-cell protein synthesis -relationship to p200 cleavage. J Virol 61:2480–2488

Macejak DG, Sarnow P (1991) Internal initiation of translation mediated by the 5′ leader of a cellular mRNA. Nature 353:90–94

Mader S, Lee H, Pause A, Sonenberg N (1995) The translation initiation factor eIF-4E binds to a common motif shared by the translation factor eIF-4gamma and the translational repressors 4E-binding proteins. Mol Cell Biol 15:4990–4997

Manzella JM, Rychlik W, Rhoads RE, Hershey JWB, Blackshear PJ (1991) Insulin induction of ornithine decarboxylase. Importance of mRNA secondary structure and phosphorylation of eucaryotic initiation factors eIF-4B and eIF-4E. J Biol Chem 266:2383–2389

Marcotrigiano J, Gingras AC, Sonenberg N, Burley SK (1997) Cocrystal structure of the messenger RNA 5′ cap-binding protein (eIF4E) bound to 7-methyl-GDP. Cell 89:951–961

Marcotrigiano J, Gingras A-C, Sonenberg N, Burley SK (1999) Cap-dependent translation initiation in eukaryotes is regulated by a molecular mimic of eIF4G. Mol Cell 3:707–716

Marissen WE, Lloyd RE (1998) Eukaryotic translation initiation factor 4G is targeted for proteolytic cleavage by caspase 3 during inhibition of translation in apoptotic cells. Mol Cell Biol 18:7565–7574

Marissen WE, Guo Y, Thomas AAM, Matts RL, Lloyd RE (2000) Identification of caspase-3-mediated cleavage and functional alteration of eukaryotic initiation factor 2α in apoptosis. J Biol Chem 275:9314–9323

Marth JD, Overell RW, Meier KE, Krebs EG, Perlmutter RM (1988) Translational activation of the lck proto-oncogene. Nature 332:171–173

Marx SO, Marks AR (1999) Cell cycle progression and proliferation despite 4BP-1 dephosphorylation. Mol Cell Biol 19:6041–6047

McKendrick L, Pain VM, Morley SJ (1999) Translation initiation factor 4E. Int J Biochem Cell Biol 31:31–35

Merrick WC (1992) Mechanism and regulation of eukaryotic protein synthesis. Microbiol Rev 56:291–315

Merrick WC, Hershey JWB (1996) The pathway and mechanism of protein synthesis. In: Hershey JWB, Mathews MB, Sonenberg N (eds) Translational control. Cold Spring Harbor Laboratory Press, Cold Spring Harbor, New York, pp 31–70

Methot N, Pause A, Sonenberg N (1994) The translation initiation factor eIF-4B contains an RNA-binding region that is distinct and independent from its ribonucleoprotein consensus sequence. Mol Cell Biol 14:2307–2316

Methot N, Song MS, Sonenberg N (1996) A region rich in aspartic acid, arginine, tyrosine and glycine (DRYG) mediates eukaryotic initiation factor 4B (eIF4B) self-association and interaction with eIF3. Mol Cell Biol 16:5328–5334

Methot N, Rom E, Olsen H, Sonenberg N (1997) The human homologue of the yeast Prt1 protein is an integral part of the eukaryotic initiation factor 3 complex and interacts with p170. J Biol Chem 272:1110–1116

Meyer K, Petersen A, Niepmann M, Beck E (1995) Interaction of eukaryotic initiation factor eIF-4B with a picornavirus internal translation initiation site. J Virol 69:2819–2824

Meyuhas O, Avni D, Shama S (1996) Translational control of ribosomal protein mRNAs in eukaryotes. In: Hershey JWB, Mathews MB, Sonenberg N (eds) Translational control. Cold Spring Harbor Laboratory Press, Plainsview, NY, pp 363–388

Milburn SC, Hershey JWB, Davies MV, Kelleher K, Kaufman RJ (1990) Cloning and expression of eukaryotic initiation factor 4B cDNA: sequence determination identifies a common RNA recognition motif. EMBO J 9:2783–2790

Minich WB, Balasta ML, Goss DJ, Rhoads RE (1994) Chromatographic resolution of in vivo phosphorylated and nonphosphorylated eukaryotic translation initiation factor eIF-4E: Increased cap affinity of the phosphorylated form. Proc Natl Acad Sci USA 91:7668–7672

Morino S, Imataka H, Svitkin YV, Pestova TV, Sonenberg N (2000) Eukaryotic translation initiation factor 4E (eIF4E) binding site and the middle one-third of eIF4GI constitute the core domain for cap-dependent translation, and the c-terminal one-third functions as a modulatory region. Mol Cell Biol 20:468–477

Morley SJ (1994) Signal transduction mechanisms in the regulation of protein synthesis. Mol Biol Rep 19:221–231

Morley SJ (1996) Regulation of components of the translational machinery by protein phosphorylation. In: Clemens MJ (ed) Protein phosphorylation in cell growth regulation. Harwood, Amsterdam, pp 197–224

Morley SJ (1997) Signalling through either the p38 or ERK mitogen-activated protein (MAP) kinase pathway is obligatory for phorbol ester and T cell receptor complex (TCR-CD3)-stimulated phosphorylation of initiation factor (eIF) 4E in Jurkat T cells. FEBS Lett 418:327–332

Morley SJ, McKendrick L (1997) Involvement of stress-activated protein kinase and p38/RK mitogen-activated protein kinase signaling pathways in the enhanced phosphorylation of initiation factor 4E in NIH 3T9 cells. J Biol Chem 272:17887–17893

Morley SJ, Pain VM (1995a) Hormone-induced meiotic maturation in Xenopus oocytes occurs independently of $p70^{s6k}$ activation and is associated with enhanced initiation factor (eIF)-4F phosphorylation and complex formation. J Cell Sci 108:1751–1760

Morley SJ, Pain VM (1995b) Translational regulation during activation of porcine peripheral blood lymphocytes: association and phosphorylation of the alpha and gamma subunits of the initiation factor complex eIF-4F. Biochem J 312:627–635

Morley SJ, Traugh JA (1990) Differential stimulation of phosphorylation of initiation factors eIF-4F, eIF-4B, eIF-3, and ribosomal protein S6 by insulin and phorbol esters. J Biol Chem 265:10611–10616

Morley SJ, Traugh JA (1993) Stimulation of translation in 3T3-L1 cells in response to insulin and phorbol ester is directly correlated with increased phosphate labelling of initiation factor (eIF-) 4F and ribosomal protein S6. Biochimie 75:985–989

Morley SJ, Rau M, Kay JE, Pain VM (1993) Increased phosphorylation of eukaryotic initiation factor 4a during early activation of T lymphocytes correlates with increased initiation factor 4F complex formation. Eur J Biochem 218:39–48

Morley SJ, Curtis PS, Pain VM (1997) eIF4G: translation's mystery factor begins to yield its secrets. RNA 3:1085–1104

Morley SJ, McKendrick L, Bushell M (1998) Cleavage of translation initiation factor 4G (eIF4G) during anti-Fas IgM-induced apoptosis does not require signalling through the p38 mitogen-activated protein (MAP) kinase. FEBS Lett 438:41–48

Morley SJ, Jeffrey I, Bushell M, Pain VM, Clemens MJ (2000) Differential requirements for caspase-8 activity in the mechanism of phosphorylation of elF2, cleavage of eIF4GI and signaling events associated with the inhibition of protein synthesis in apoptotic Jurkat T cells. FEBS Lett 477:229–236

Muckenthaler M, Gray NK, Hentze MW (1998) IRP-1 binding to ferritin mRNA prevents the recruitment of the small ribosomal subunit by the cap-binding complex eIF4F. Mol Cell 2:383–388

Nagata S (1997) Apoptosis by death factor. Cell 88:355–365

Naranda T, Strong WB, Menaya J, Fabbri BJ, Hershey JWB (1994) Two structural domains of initiation factor eIF-4B are involved in binding to RNA. J Biol Chem 269:14465–14472

Nathan CA, Carter P, Liu L, Li BD, Abreo F, Tudor A, Zimmer SG, De Benedetti A (1997a) Elevated expression of eIF4E and FGF-2 isoforms during vascularization of breast carcinomas. Oncogene 15:1087–1094

Nathan CAO, Liu L, Li BD, Abreo FW, Nandy I, De Benedetti A (1997b) Detection of the proto-oncogene eIF4E in surgical margins may predict recurrence in head and neck cancer. Oncogene 15:579–584

Negulescu D, Leong LEC, Chandy KG, Semler BL, Gutman GA (1998) Translation initiation of a cardiac voltage-gated potassium channel by internal ribosome entry. J Biol Chem 273:20109–20113

Ohlmann T, Rau M, Morley SJ, Pain VM (1995) Proteolytic cleavage of initiation factor eIF-4gamma in the reticulocyte lysate inhibits translation of capped mRNAs but enhances that of uncapped mRNAs. Nucleic Acids Res 23:334–340

Ohlmann T, Rau M, Pain VM, Morley SJ (1996) The C-terminal domain of eukaryotic protein synthesis initiation factor (eIF) 4G is sufficient to support cap-independent translation in the absence of eIF4E. EMBO J 15:1371–1382

Ohlmann T, Pain VM, Wood W, Rau M, Morley SJ (1997) The proteolytic cleavage of eukaryotic initiation factor (eIF) 4G is prevented by eIF4E binding protein (PHAS-I;4E- BP1) in the reticulocyte lysate. EMBO J 16:844–855

Oumard A, Hennecke M, Hauser H, Nourbakhsh M (2000) Translation of NRF mRNA is medaited by highly efficient ribosome entry. Mol Cell Biol 20:2755–2759

Pain VM (1996) Initiation of protein synthesis in eukaryotic cells. Eur J Biochem 236:747–771

Panniers R, Stewart E, Merrick WC, Henshaw EC (1985) Mechanism of inhibition of polypeptide chain initiation in heat-shocked Ehrlich cells involves reduction of eukaryotic initiation factor 4F activity. J Biol Chem 260:9648–9653

Paraskeva E, Gray NK, Schlaeger B, Wehr K, Hentze MW (1999) Ribosomal pausing and scanning arrest as mechanisms of translational regulation from cap-distal iron-responsive elements. Mol Cell Biol 19:807–816

Pause A, Sonenberg N (1992) Mutational analysis of a DEAD box RNA helicase: the mammalian translation initiation factor eIF-4A. EMBO J 11:2643–2654

Pause A, Sonenberg N (1993) Helicases and RNA unwinding in translation. Curr Opin Struct Biol 3:953–959

Pause A, Belsham GJ, Gingras A-C, Donzé O, Lin T-A, Lawrence JC Jr, Sonenberg N (1994a) Insulin-dependent stimulation of protein synthesis by phosphorylation of a regulator of 5′-cap function. Nature 371:762–767

Pause A, Méthot N, Svitkin Y, Merrick WC, Sonenberg N (1994b) Dominant negative mutants of mammalian translation initiation factor eIF-4A define a critical role for eIF-4F in cap-dependent and cap-independent initiation of translation. EMBO J 13:1205–1215

Pestova TV, Hellen CUT (1999) Ribosome recruitment and scanning: what's new? Trends Biochem Sci 24:85–87

Pestova TV, Shatsky IN, Hellen CUT (1996). Functional dissection of eukaryotic initiation factor 4F: The 4A subunit and the central domain of the 4G subunit are sufficient to mediate internal entry of 43 S preinitiation complexes. Mol Cell Biol 16:6870–6878.

Pestova TV, Borukhov SI, Hellen CUT (1998) Eukaryotic ribosomes require initiation factors 1 and 1 A to locate initiation codons. Nature 394:854–859

Pestova TV, Lomakin IB, Choi SK, Dever TE, Hellen CUT (2000) The joining of ribosomal subunits in eukaryotes requires eIF5B. Nature 403:332–335

Peter ME, Heufelder AE, Hengartner MO (1997) Advances in apoptosis research. Proc Natl Acad Sci USA 94:12736–12737

Piron M, Vende P, Cohen J, Poncet D (1998) Rotavirus RNA-binding protein NSP3 interacts with eIF4GI and evicts the poly(A) binding protein from eIF4F. EMBO J 17:5811–5821

Porter AG (1999) Protein translocation in apoptosis. Trends Cell Biol 9:394–401

Poulin F, Gingras AC, Olsen H, Chevalier S, Sonenberg N (1998) 4E-BP3, a new member of the eukaryotic initiation factor 4E-kinding protein family. J Biol Chem 273:14002–14007

Pozner A, Goldenberg D, Negreanu V, Le SY, Elroy-Stein O, Levanon D, Groner Y (2000) Transcription-coupled translation control of AML1/RUNX1 is mediated by cap-and internal ribsome entry site-dependent mechanisms. Mol Cell Biol 20:2297–2307

Preiss T, Hentze MW (1999) From factors to mechanisms: translation and translational control in eukaryotes. Curr Opin Genet Dev 9:515–521

Proud CG, Denton RM (1998) Molecular mechanisms for the control of translation by insulin. Biochem J 328:329–341

Ptushkina M, Von der Haar T, Karim MM, McCarthy JEG (1999) Repressor binding to a dorsal regulatory site traps human eIF4E in a high cap-affinity state. EMBO J 18:4068–4075

Pyronnet S, Imataka H, Gingras A-C, Fukunaga R, Hunter T, Sonenberg N (1999) Human eukaryotic translation initiation factor 4G (eIF4G) recruits Mnk1 to phosphorylate eIF4E. EMBO J 18:270–279

Pyronnet S, Vagner S, Bouisson M, Prats AC, Vaysse N, Pradayrol L (1996) Relief of ornithine decarboxylase messenger RNA translational repression induced by alternative splicing of its 5′ untranslated region. Cancer Res 56:1742–1745

Rathmell JC, Thompson CB (1999) The central effectors of cell death in the immune system. Annu Rev Immunol 17:781–828

Rau M, Ohlmann T, Morley SJ, Pain VM (1996) A re-evaluation of the cap-binding protein, eIF4E, as a rate-limiting factor for initiation of translation in reticulocyte lysate. J Biol Chem 271:8983–8990

Raught B, Gingras A-C, Gygi SP, Imataka H, Morino S, Gradi A, Aebersold R, Sonenberg N (2000) Serum-stimulated, rapamycin-sensitive phosphorylation sites in the eukaryotic initiation factor 4GI. EMBO J 19:434–444

Ray BK, Lawson TG, Kramer JC, Cladaras MH, Grifo JA, Abramson RD, Merrick WC, Thach RE (1985) ATP-dependent unwinding of messenger RNA structure by eukaryotic initiation factors. J Biol Chem 260:7651–7658

Rhoads RE (1993) Regulation of eukaryotic protein synthesis by initiation factors. J Biol Chem 268:3017–3020

Rhoads RE (1999) Signal transduction pathways that regulate eukaryotic protein synthesis. J Biol Chem 274:30337–30340

Richter-Cook NJ, Dever TE, Hensold JO, Merrick WC (1998) Purification and characterization of a new eukaryotic protein translation factor – eukaryotic initiation factor 4H. J Biol Chem 273: 7579–7587

Rinker-Schaeffer CW, Graff JR, De Benedetti A, Zimmer SG, Rhoads RE (1993) Decreasing the level of translation initiation factor 4E with antisense RNA causes reversal of ras-mediated transformation and tumorigenesis of cloned rat embryo fibroblasts. Int J Cancer 55:841–847

Roberts LO, Seamons RA, Belsham GJ (1998) Recognition of picornavirus internal ribosome entry sites within cells; influence of cellular and viral proteins. RNA 4:520–529

Rogers GW Jr, Richter NJ, Merrick WC (1999) Biochemical and kinetic characterization of the RNA helicase activity of eukaryotic initiation factor 4A. J Biol Chem 274:12236–12244

Rosenwald IB, Lazaris-Karatzas A, Sonenberg N, Schmidt EV (1993) Elevated levels of cyclin D1 protein in response to increased expression of eukaryotic initiation factor 4E. Mol Cell Biol 13:7358–7363

Rosenwald IB, Kaspar R, Rousseau D, Gehrke L, Leboulch P, Chen JJ, Schmidt EV, Sonenberg N, London IM (1995) Eukaryotic translation initiation factor 4E regulates expression of cyclin D1 at transcriptional and post-transcriptional levels. J Biol Chem 270:21176–21180

Rosenwald IB, Chen JJ, Wang ST, Savas L, London IM, Pullman J (1999) Upregulation of protein synthesis initiation factor eIF-4E is an early event during colon carcinogenesis. Oncogene 18:2507–2517

Rothe M, Pan MG, Henzel WJ, Ayres TM, Goeddel DV (1995) The TNFR2-TRAF signaling complex contains two novel proteins related to baculoviral inhibitor of apoptosis proteins. Cell 83:1243–1252

Rousseau D, Gingras A-C, Pause A, Sonenberg N (1996a) The eIF4E-binding protein-1 and protein-2 are negative regulators of cell growth. Oncogene 13:2415–2420

Rousseau D, Kaspar R, Rosenwald I, Gehrke L, Sonenberg N (1996b) Translation initiation of ornithine decarboxylase and nucleocytoplasmic transport of cyclin D1 mRNA are increased in cells overexpressing eukaryotic initiation factor 4E. Proc Natl Acad Sci USA 93:1065–1070

Rowan S, Fisher DE (1997) Mechanisms of apoptotic cell death. Leukemia 11:457–465

Rozen F, Edery I, Meerovitch K, Dever TE, Merrick WC, Sonenberg N (1990) Bidirectional RNA helicase activity of eukaryotic initiation factors 4A and 4F. Mol Cell Biol 10:1134–1144

Rust RC, Ochs K, Meyer K, Beck E, Niepmann M (1999) Interaction of eukaryotic initiation factor eIF4B with the internal ribosome entry site of foot-and-mouth disease virus is independent of the polypyrimidine tract-binding protein. J Virol 73:6111–6113

Rychlik W, Russ MS, Rhoads RE (1987) Phosphorylation site of eukaryotic initiation factor 4E. J Biol Chem 262:10434–10437

Sachs AB, Buratowski S (1997) Common themes in translational and transcriptional regulation. Trends Biochem Sci 22:189–192

Sachs AB, Davis RW, Kornberg RD (1987) A single domain of yeast poly(A)-binding protein is necessary and sufficient for RNA-binding and cell viability. Mol Cell Biol 7:3268–3276

Satoh S, Hijikata M, Handa H, Shimotohno K (1999) Caspase-mediated cleavage of eukaryotic translation initiation factor subunit 2a. Biochem J 342:65–70

Scheper GC, Mulder J, Kleijn M, Voorma HO, Thomas AAM, Van Wijk R (1997) Inactivation of eIF2B and phosphorylation of PHAS-I in heat-shocked rat hepatoma cells. J Biol Chem 272:26850–26856

Schiavi A, Hudder A, Werner R (1999) Connexin43 mRNA contains a functional internal ribosome entry site. FEBS Lett 464:118–122
Schulze-Osthoff K, Ferrari D, Los M, Wesselborg S, Peter ME (1998) Apoptosis signaling by death receptors. Eur J Biochem 254:439–459
Sella O, Gerlitz G, Le SY, Elroy-Stein O (1999) Differentiation-induced internal translation of c-sis mRNA: Analysis of the cis elements and their differentiation- linked binding to the hnRNP C protein. Mol Cell Biol 19:5429–5440
Shah OJ, Antonetti DA, Kimball SR, Jefferson LS (1999) Leucine, glutamine and tyrosine reciprocally modulate the translation initiation factors eIF4F and eIF2B in perfused rat liver. J Biol Chem 274:36168–36175
Shah OJ, Kimball SR, Jefferson LS (2000) Acute attenuation of translation initiation and protein saynthesis by glucocorticoids in skeletal muscle. Am J Physiol 278:E76–E82
Shantz LM, Coleman CS, Pegg AE (1996) Expression of an ornithine decarboxylase dominant-negative mutant reverses eukaryotic initiation factor 4E-induced cell transformation. Cancer Res 56:5136–5140
Sizova DV, Kolupaeva VG, Pestova TV, Shatsky IN, Hellen CUT (1998) Specific interaction of eukaryotic translation initiation factor 3 with the 5′ nontranslated regions of hepatitis C virus and classical swine fever virus RNAs. J Virol 72:4775–4782
Sonenberg N (1994) Regulation of translation and cell growth by eIF-4E. Biochimie 76:839–846
Sonenberg N (1996) mRNA 5′ cap-binding protein eIF4E and control of cell growth. In: Hershey JWB, Mathews MB, Sonenberg N (eds) Translational control. Cold Spring Harbor Laboratory Press, Plainsview, NY, pp 245–269
Stein I, Itin A, Einat P, Skaliter R, Grossman Z, Keshet E (1998) Translation of vascular endothelial growth factor mRNA by internal ribosome entry: Implications for translation under hypoxia. Mol Cell Biol 18:3112–3119
Stennicke HR, Salvesen GS (1999) Catalytic properties of the caspases. Cell Death Differ 6:1054–1059
Stoneley M, Paulin FEM, Le Quesne JPC, Chappell SA, Willis AE (1998) C-Myc 5′ untranslated region contains an internal ribosome entry segment. Oncogene 16:423–428
Stoneley M, Chappell SA, Jopling CL, Dickens M, MacFarlane M, Willis AE (2000a) c-myc protein synthesis is initiated from the internal ribosome entry segment during apoptosis. Mol Cell Biol 20:1162–1169
Stoneley M, Subkhankulova T, Le Quesne JPC, Coldwell M, Jopling CL, Belsham G, Willis AE (2000b) Analysis of the c-myc IRES; a potential role for cell type specific trans-acting factors and the nuclear compartment. Nucleic Acids Res 28:687–694
Svitkin YV, Gradi A, Imataka H, Morino S, Sonenberg N (1999) Eukaryotic initiation factor 4GII (eIF4GII), but not eIF4GI, cleavage correlates with inhibition of host cell protein synthesis after human rhinovirus infection. J Virol 73:3467–3472
Tarun SZ, Sachs AB (1996) Association of the yeast poly(A) tail binding protein with translation initiation factor eIF-4G. EMBO J 15:7168–7177
Tarun SZ, Wells SE, Deardorff JA, Sachs AB (1997) Translation initiation factor eIF4G mediates in vitro poly(A) tail-dependent translation. Proc Natl Acad Sci USA 94:9046–9051
Teerink H, Voorma HO, Thomas AAM (1995) The human insulin-like growth factor II leader 1 contains an internal ribosomal entry site. Biochim Biophys Acta 1264:403–408
Terada N, Franklin RA, Lucas JJ, Blenis J, Gelfand EW (1993) Failure of rapamycin to block proliferation once resting cells have entered the cell cycle despite inactivation of p70 S6 kinase. J Biol Chem 268:12062–12068
Terada N, Takase K, Papst P, Nairn AC, Gelfand EW (1995) Rapamycin inhibits ribosomal protein synthesis and induces G1 prolongation in mitogen-activated T lymphocytes. J Immunol 155:3418–3426
Tinton SA, Buc-Calderon PM (1999) Hypoxia increases the association of 4E-binding protein 1 with the initiation factor 4E in isolated rat hepatocytes. FEBS Lett 446:55–59
Tuxworth WJ, Wada H, Ishibashi Y, McDermott PJ (1999) Role of load in regulating eIF4F complex formation in adult feline cardiocytes. Am J Physiol 277:H1273–H1282

Vagner S, Gensac M-C, Maret A, Bayard F, Amalric F, Prats H, Prats A-C (1995) Alternative translation of human fibroblast growth factor 2 mRNA occurs by internal entry of ribosomes. Mol Cell Biol 15:35–44

Van der Velden AW, Thomas AAM (1999) The role of the 5′ untranslated region of an mRNA in translation regulation during development. Int J Biochem Cell Biol 31:87–106

Vary TC, Jefferson LS, Kimball SR (1999) Amino acid-induced stimulation of translation initiation in rat skeletal muscle. Am J Physiol 277:E1077–E1086

Vary TC, Jefferson LS, Kimball SR (2000) Role of eIF4E in stimulation of protein synthesis by IGF-1 in perfused rat skeletal muscle. Am J Physiol Endocrinol Metab 278:E58–E64

Vries RGJ, Flynn A, Patel JC, Wang XM, Denton RM, Proud CG (1997) Heat shock increases the association of binding protein-1 with initiation factor 4E. J Biol Chem 272:32779–32784

Wada H, Ivester CT, Carabello BA, Cooper G, McDermott PJ (1996) Translational initiation factor eIF-4E – a link between cardiac load and protein synthesis. J Biol Chem 271:8359–8364

Wallach D, Varfolomeev EE, Malinin NL, Goltsev YV, Kovalenko AV, Boldin MP (1999) Tumor necrosis factor receptor and Fas signaling mechanisms. Annu Rev Immunol 17:331–367

Wang XM, Proud CG (1997) p70 S6 kinase is activated by sodium arsenite in adult rat cardiomyocytes: Roles for phosphatidylinositol 3-kinase and p38 MAP kinase. Biochem Biophys Res Commun 238:207–212

Wang XM, Flynn A, Waskiewicz AJ, Webb BLJ, Vries RG, Baines IA, Cooper JA, Proud CG (1998) The phosphorylation of eukaryotic initiation factor eIF4E in response to phorbol esters, cell stresses, and cytokines is mediated by distinct MAP kinase pathways. J Biol Chem 273:9373–9377

Waskiewicz AJ, Flynn A, Proud CG, Cooper JA (1997) Mitogen-activated protein kinases activate the serine/threonine kinases Mnk1 and Mnk2. EMBO J 16:1909–1920

Waskiewicz AJ, Johnson JC, Penn B, Mahalingam M, Kimball SR, Cooper JA (1999) Phosphorylation of the cap-binding protein eukaryotic translation initiation factor 4E by protein kinase Mnk1 in vivo. Mol Cell Biol 19:1871–1880

Weinstein DC, Honoré E, Hemmati-Brivanlou A (1997) Epidermal induction and inhibition of neural fate by translation initiation factor 4AIII. Development 124:4235–4242

Wells SE, Hillner PE, Vale RD, Sachs AB (1998) Circularization of mRNA by eukaryotic translation factors. Mol Cell 2:135–140

Whalen SG, Gingras AC, Amankwa L, Mader S, Branton PE, Aebersold R, Sonenberg N (1996) Phosphorylation of eIF-4E on serine 209 by protein kinase C is inhibited by the translational repressors, 4E-binding proteins. J Biol Chem 271:11831–11837

Wickens M, Anderson P, Jackson RJ (1997) Life and death in the cytoplasm: Messages from the 3′ end. Curr Opin Genet Dev 7:220–232

Widmann C, Gibson S, Johnson GL (1998) Caspase-dependent cleavage of signaling proteins during apoptosis – a turn-off mechanism for anti-apoptotic signals. J Biol Chem 273: 7141–7147

Williams-Hill DM, Duncan RF, Nielsen PJ, Tahara SM (1997) Differential expression of the murine eukaryotic translation initiation factor isogenes $eIF4A_I$ and $eIF4A_{II}$ is dependent upon cellular growth status. Arch Biochem Biophys 338:111–120

Willis AE (1999) Translational control of growth factor and proto-oncogene expression. Int J Biochem Cell Biol 31:73–86

Winzen R, Kracht M, Ritter B, Wilhelm A, Chen C-YA, Shyu A-B, Muller M, Gaestel M, Resch K, Holtmann H (1999) The p38 MAP kinase pathway signals for cytokine-induced mRNA stabilisation via MAP kinase-activated protein kinase 2 and an AU-rich region-targeted mechanism. EMBO J 18:4969–4980

Wolf BB, Green DR (1999) Suicidal tendencies: apoptotic cell death by caspase family proteinases. J Biol Chem 274:20049–20052

Wyllie AH (1997) Apoptosis and carcinogenesis. Eur J Cell Biol 73:189–197

Yang DQ, Brunn GJ, Lawrence JC Jr (1999) Mutational analysis of sites in the translational regulator, PHAS-I, that are selectively phosphorylated by mTOR. FEBS Lett 453:387–390

Yoder-Hill J, Pause A, Sonenberg N, Merrick WC (1993) The p46 subunit of eukaryotic initiation factor (eIF)-4F exchanges with eIF-4A. J Biol Chem 268:5566–5573

Yoshizawa F, Kimball SR, Jefferson LS (1997) Modulation of translation initiation in rat skeletal muscle and liver in response to food intake. Biochem Biophys Res Commun 240:825–831

Yueh A, Schneider RJ (1996) Selective translation initiation by ribosome jumping in adenovirus-infected and heat-shocked cells. Genes Dev 10:1557–1567

Zhou BB, Li HL, Yuan JY, Kirschner MW (1998) Caspase-dependent activation of cyclin-dependent kinases during fas-induced apoptosis in Jurkat cells. Proc Natl Acad Sci USA 95:6785–6790

Regulation of the Activity of Eukaryotic Initiation Factors in Stressed Cells

Gert C. Scheper, Roel Van Wijk, and Adri A.M. Thomas[1]

1 Stress and Protein Synthesis

1.1 Inhibition of General Protein Synthesis

Exposure of cells to elevated temperatures, also known as heat shock, or exposure to other kinds of environmental stress, gives rise to a general inhibition of protein synthesis, on the one hand, and to the increased synthesis of a group of chaperone proteins, on the other hand (Rhoads and Lamphear 1995; Duncan 1996). These chaperone proteins or heat shock proteins (HSPs) function mainly in folding and refolding of denatured or newly synthesized proteins and therefore counteract the toxic effects of denatured proteins on the cell (Wu 1995). The inhibition of protein synthesis is achieved by a reduced initiation of translation through the inactivation of various eukaryotic initiation factors (eIFs; Duncan 1996). Most studies have implicated the inactivation of eIF2, which is involved in the delivery of the initiator methionyl tRNA to the small ribosomal subunit, and the inactivation of eIF4E, a key component in the binding of several initiation factors to the mRNA, as the main causes of the reduced levels of initiation (e.g., Duncan and Hershey 1989; Duncan et al. 1995; Feigenblum and Schneider 1996; Scheper et al. 1997; Wang et al. 1998)

1.2 Phosphorylation of eIF2

eIF2 delivers the initiator tRNA in a GTP-bound state to the 40 S small ribosomal subunit. At the joining of the 60 S ribosomal subunit to the 40 S subunit, this GTP is hydrolyzed to GDP and eIF2 is released in an inactive state (Pain 1996). The guanine nucleotide exchange activity of the five-subunit complex eIF2B ensures the recycling of GDP for GTP, rendering eIF2 active again (Panniers and Henshaw 1983; Cigan et al. 1993; Price and Proud 1994). The

[1] Department of Molecular Cell Biology, Utrecht University, Utrecht, The Netherlands
Present address: G.C. Scheper, Department of Anatomy and Physiology, University of Dundee, Dundee DD1 5EH, Scotland

Progress in Molecular and Subcellular Biology, Vol. 27
Robert E. Rhoads (Ed.)

main mechanism by which eIF2 can be inactivated is by phosphorylation of its α subunit at Serine 51 (Price et al. 1991; Proud 1992; Kimball et al. 1998). Phosphorylation of eIF2 at this position leads to the formation of a stable complex between eIF2 and eIF2B, thereby reducing the levels of active eIF2B, and consequently the levels of eIF2-GTP in the cell.

Four main types of eIF2 kinases have been cloned and characterized so far: HRI, GCN2, PKR, and PERK. HRI is an eIF2α kinase that is activated by various kinds of stress, such as heat shock, heme deficiency, and oxidative stress (Matts et al. 1992, 1993; Chen et al. 1994; Chen and London 1995; Uma et al. 1997; Xu et al. 1997; Berlanga et al. 1998; Thulasiraman et al. 1998). Its activation is regulated in several ways through interaction with heat shock proteins, especially HSP90 and HSC70. HSP90 is required for the proper folding of the kinase, but does not seem to play a role in the eventual transformation of HRI into an active kinase. HSC70 is required, on the one hand, for the transformation of HRI into an active kinase, but subsequently HSC70 diminishes the activity of transformed HRI (Thulasiraman et al. 1998; Uma et al. 1999). Activation of HRI is therefore likely to be regulated by stress conditions that will lead to the formation of denatured proteins, to which HSC70 will bind. This would release the inhibitory effect of HSC70 on HRI, rendering the kinase active. Although HRI seems to be a good candidate for the stress-activated eIF2 kinase, one has to keep in mind that this kinase is only present in cells of erythroid origin.

GCN2 is an eIF2α kinase that is activated upon amino acid starvation of yeast cells, leading to the specific expression of GCN4, a transcriptional activator of factors involved in amino acid synthesis. Very recently, the mammalian homologue of GCN2 has been identified (Berlanga et al. 1999; Sood et al. 2000). So far, the data suggest that this kinase is switched on only upon amino acid starvation, and a possible role in the stress-induced phosphorylation of eIF2α seems unlikely.

PEK, or PERK (PKR-like endoplasmic reticulum kinase), has recently been identified as an endoplasmic reticulum resident eIF2α kinase that is specifically activated by ER stresses, such as treatment of cells with thapsigargin and tunicamycin. Stresses like heat shock, arsenite treatment and UV light did not activate PERK (Shi et al. 1998, 1999; Harding et al. 1999). Further studies will shed more light on the importance of this kinase for eIF2α phosphorylation and the regulation of protein synthesis under various conditions.

PKR is the best-studied eIF2α kinase. This kinase that is activated by double-stranded RNA plays a pivotal role in the interferon-induced antiviral and antiproliferative responses upon viral infection of cells. Many viruses have in their turn developed various ways to escape from the antiviral action of PKR activation and eIF2α phosphorylation. For example, vaccinia virus encodes two different proteins, E3L that sequesters dsRNA, and K3L that mimics eIF2α and acts as a pseudosubstrate for PKR (Davies et al. 1992, 1993; Sharp et al. 1998). A completely different mechanism is employed by herpes simplex virus, which produces a phosphatase regulatory subunit that specifically targets phosphatase 1 to eIF2 (He et al. 1997), while adenovirus and Epstein Barr virus-associated RNAs prevent the activation of PKR (Clemens et al. 1994). PKR is involved in cell growth and a tumour-suppressor activity of this kinase has been suggested (e.g., Meurs et al. 1993; reviewed in Clemens 1997). PKR activation also occurs under stress conditions (Srivastava et al. 1995; Brostrom et al. 1996), although by which means has been unclear for several years. Two mechanisms involved in the regulation of PKR activity, separate from activation by dsRNA, could explain the stress-induced activation of PKR. One of these mechanisms involves a protein called $P58^{IPK}$ that appears to keep PKR in an inactive state. Overexpression of $P58^{IPK}$ causes malignant transformation by reducing PKR activity in the cell (Barber et al. 1994). $P58^{IPK}$ seems to be a member of the family of heat shock proteins and its binding to PKR protein is in turn regulated by at least two proteins, HSP40 (Melville et al. 1997) and $P52^{rIPK}$ (Gale et al. 1998). A part of the latter protein shows some homology to a charged domain of HSP90. Therefore, as for HRI, the activity of PKR appears to be dependent on the activity of various HSPs or related proteins.

The second mechanism by which PKR can be activated is through interaction with RAX and PACT, two PKR-associated proteins (Patel and Sen 1998; Ito et al. 1999). As most stress conditions will not rapidly increase the levels of double-stranded RNA, the discovery of these activators could explain the dsRNA-independent activation of PKR, which is seen under most stress conditions. In vivo, RAX activity seems to be regulated by phosphorylation and this phosphorylation precedes activation of PKR. However, the signalling pathways leading to RAX phosphorylation and its exact role in the activation of PKR remain to be clarified.

1.3
Inactivation of eIF2B

Although phosphorylation of eIF2α and, therefore, inactivation of eIF2B can explain the inhibition of protein synthesis found in stressed cells in most cases, there is a slight discrepancy between the two phenomena. At mild heat shock temperatures (40–41 °C), the increased phosphorylation of eIF2α was not observed (Scheper et al. 1997), whereas at these temperatures, protein synthesis was greatly compromised. In Tables 1 and 2 we have summarized our data on the levels of protein synthesis and eIF2B activity following heat shock or arsenite treatment of rat hepatoma H35 cells.

Heat shock as well as arsenite treatment caused a severe inhibition of protein synthesis (Table 1). As expected, the effect on protein synthesis was more severe at higher temperatures or higher arsenite concentrations. The initial effects of arsenite were somewhat more eminent than the effects of the chosen temperatures, but cells could recover from the arsenite treatments more efficiently than from the exposure to high temperatures. This reflects differences in the toxic effects of the two different stress treatments.

Table 1. Protein synthesis in stressed cells

Treatment	Protein synthesis (%) after treatment	Protein synthesis (%) after recovery
Control	100	100
30 min at 42.5 °C	26	75
30 min at 43.5 °C	21	44
30 min at 44.5 °C	8	16
1 h 50 μM As	19	93
1 h 125 μM As	14	59
1 h 300 μM As	4	49

H35 rat hepatoma cells were treated as indicated in the left column. After the stress treatments the cells were washed twice with PBS. Either methionine-free medium with [^{35}S]-methionine/cysteine was added for 30 min (*middle column*) to measure [^{35}S]-methionine/cysteine incorporation, or normal medium was added and the cells were left at 37 °C for 5 h to recover from the stress conditions. After this recovery period the cells were labelled with [^{35}S]-methionine/cysteine in methionine-free medium for 1 h (*right column*). Protein synthesis was measured as described in experimental procedures. Control cells were kept at 37 °C for the duration of the experiment and the values found for [^{35}S]-methionine/cysteine incorporation in the control cells were set at 100% (corresponding to 2.2×10^6 and 4.8×10^6 cpm for the whole sample).

Table 2. Inhibition of eIF2B activity in arsenite-treated cells

Arsenite concentration (μM)	eIF2B activity in stressed cells (%)	eIF2B activity in recovered cells (%)
0	100	100
50	43	138
125	45	185
300	7.5	158

H35 cells were incubated in the presence of various concentrations of sodium arsenite and harvested as described in experimental procedures. eIF2B activity was measured directly after 1 h of treatment or after recovery of the cells for 5 h in medium lacking arsenite. The eIF2B activity of untreated cells was set at 100% (corresponding to the exchange of approximately 40% of the [^{3}H]-labelled GDP for GTP).

eIF2B activity was determined in these cells (Table 2). Inhibition of this factor was shown to be the major cause of inhibition of protein synthesis under mild heat shock conditions (Scheper et al. 1997) and eIF2B might also be compromised by arsenite treatment. Arsenite treatment of H35 cells led to a severe inhibition of eIF2B activity (Table 2). Only 7.5% activity of control levels was found after treatment at the highest concentration. Interestingly, the inactivation of eIF2B was rapidly reversed after recovery of the treated cells. In the recovered cells even higher activities for eIF2B were found. In contrast, eIF2B activity in heat-shocked cells was still compromised after several hours of recovery (Scheper et al. 1997). This difference in apparent toxicity of the two treatments was also reflected in the less severe effect of arsenite on protein synthesis in recovering cells (Table 1).

The way in which eIF2B was inactivated remains to be established, although phosphorylation of the α subunit of eIF2 seems a likely cause (Duncan and Hershey 1987; Brostrom et al. 1996). Interestingly, arsenite activates the rapamycin-dependent signalling pathway that signals through PI3K and FRAP (FK506-binding protein-rapamycin-associated protein) (Wang and Proud 1997), and eIF2B can be regulated via similar pathways (Welsh et al. 1997b; Kleijn et al. 1998b). The increased eIF4E binding protein-1 (4E-BP1) phosphorylation in heat-shocked H35 cells (Scheper et al. 1997; see also below in Fig. 2) was also sensitive to rapamycin (not shown), indicating that stress activates the rapamycin-dependent signalling pathway in these cells.

It is clear that eIF2B activity is closely related with the rapid inhibition of protein synthesis. Also mild heat shock treatments that severely inhibit protein synthesis in these cells largely reduce eIF2B activity (Scheper et al. 1997). Under these conditions eIF2α phosphorylation was not changed and the cause for reduced protein synthesis is probably due to decreased eIF2B activity. We have tried unsuccessfully to restore eIF2B activity by adding NADPH and ATP, which have been reported to bind to and activate eIF2B (Dholakia et al. 1986, 1989). Another possibility is the inactivation of eIF2B by phosphorylation, as phosphorylation of the eIF2B ε subunit at Ser540 is known to cause reduced eIF2B activity (Welsh and Proud 1993; Welsh et al. 1997a). This residue is phosphorylated by glycogen-synthase kinase 3, but at present no evidence exists that it is regulated by stress. A third possibility is that the activity of eIF2B is directly regulated by interaction with HSPs. The activity of eIF2B is very sensitive to heat treatment in vitro, and such a treatment leads to the rapid denaturation of the five-subunit complex (Scheper et al. 1998). Hypothetically, this observation indicates another involvement of the heat shock proteins, which can protect against denaturation, in the regulation of eIF2 activity. The reduced eIF2B activity in arsenite-treated cells is most likely caused by phosphorylation of eIF2α (Duncan and Hershey 1987; Brostrom et al. 1996).

1.4 Inactivation of the Cap-Binding Complex eIF4F

Numerous studies have addressed the possible role of reduced eIF4E activity in the inhibition of protein synthesis following heat treatment of cells (Duncan and Hershey 1989; Lamphear and Panniers 1990; Duncan et al. 1995; Rhoads and Lamphear 1995). This protein binds the cap structure that is present at the 5′ end of all eukaryotic cellular mRNAs and thereby selects messengers that will be translated. The activity of eIF4E can be regulated in at least three ways: (1) by phosphorylation at Ser209 (Flynn and Proud 1995; Joshi et al. 1995) the cap-binding activity of eIF4E is enhanced (Minich et al. 1994). Phosphorylated eIF4E is preferentially found in 48 S initiation complexes (Hiremath et al. 1989). (2) The availability of eIF4E is regulated through binding to 4E-BPs. Phosphorylation of the 4E-BPs leads to dissociation from eIF4E, thereby enhancing the amount of eIF4E available for translation initiation (Lin et al. 1994; Pause et al. 1994). (3) The activity of eIF4E is likely to be regulated by its interaction with eIF4G (Haghighat and Sonenberg 1997). This interaction leads to the formation of a cap-binding complex eIF4F, in which the RNA helicase eIF4A is also present. eIF4G is a scaffolding protein enabling the interaction of various translation initiation factors and the ribosome (Lamphear et al. 1995; Mader et al. 1995; Tarun and Sachs 1996; Ptushkina et al. 1999; Morino et al. 2000).

Interestingly, the association of eIF4E with eIF4G brings eIF4E in the proximity of Mnk1. This kinase, which is regulated by the mitogen-activated protein kinases (MAPKs or ERKs), as well as by the stress-activated p38 MAP kinases (Fukunaga and Hunter 1997; Waskiewicz et al. 1997), is associated with eIF4G (Pyronnet et al. 1999; Waskiewicz et al. 1999). Mnk1 has been shown to phosphorylate eIF4E at Ser209 (Waskiewicz et al. 1997), and is now thought to be the physiological kinase of eIF4E.

1.5 Dephosphorylation of eIF4E

Based on results obtained with heat-shocked cells, dephosphorylation of eIF4E upon stress treatment of cells has been the rule for approximately 10 years. However, by using other kinds of stress and several kinds of cell types, this rule does not seem to be as general as accepted. Several reports have recently shown some unexpected stress-induced changes in factors involved in cap binding. The protein synthesis inhibitor anisomycin induced eIF4E phosphorylation (Morley 1997) and arsenite induced both eIF4E phosphorylation and 4E-BP1 phosphorylation (Wang and Proud 1997; Figs. 2 and 3). These events are thought to increase protein synthesis, but similar to heat shock, arsenite causes a severe inhibition of protein synthesis (Table 1). Experiments in *Xenopus* kidney cell lines have shown that heat shock can even cause increased eIF4E phosphorylation (Fraser and Morley 1997). All these data taken together clearly show that inactivation of eIF4E is not the cause for the stress-induced inhibition of protein synthesis, although dephosphorylation of eIF4E and increased binding to 4E-BPs might have an additional effect under some conditions.

1.6 Binding of eIF4E to 4E-BPs

Various reports have shown the increased binding of eIF4E to its binding proteins upon heat shock, which leads to reduced levels of eIF4E available for eIF4F formation (Flynn et al. 1997; Vries et al. 1997; Wang et al. 1998). However, this response is clearly stress- and cell-type dependent; e.g., arsenite induces phosphorylation of 4E-BP1 (Vries et al. 1997; Wang and Proud 1997) and, in contrast to the reports mentioned above, heat shock can also cause decreased binding of 4E-BP1 to eIF4E in certain cell types (Scheper et al. 1997). Phosphorylation of 4E-BP1 and release from eIF4E upon stimulation of cells have been reviewed (Kleijn et al. 1998a).

2
Preferential Translation of HSP mRNAs

As mentioned above, dephosphorylation of eIF4E has been found upon heat shock of cells in several cases (Duncan and Hershey 1989; Duncan et al. 1995; Scheper et al. 1997). This has led to the idea that the HSP mRNAs have a high affinity for eIF4E and therefore can be translated under heat shock conditions when the activity of eIF4E is decreased (Rhoads and Lamphear 1995; Duncan 1996). Although the ability to be translated under stress conditions was shown to reside in the 5′ untranslated region (5′ UTR) of the HSP mRNAs (Klemenz et al. 1985; McGarry and Lindquist 1985; Lindquist and Petersen 1990), this is probably only true for some of the HSP mRNAs. A survey of all known sequences for the 5′ UTRs of the HSPs revealed that, in general, these messengers have the same characteristics as the 5′ UTRs of a control group of normal cellular mRNAs (Joshi and Nguyen 1995).

Very recently, Yueh and Schneider have reported that the human HSP70 mRNA can be translated by a mechanism that is known as ribosomal shunting (Yueh and Schneider 2000). In this process, the ribosome is thought to bind to the 5′ cap structure and is then able to translocate to the initiation codon without scanning of the intervening 5′ UTR sequence. This mechanism has been described in most detail for two viral RNAs: the 35 S RNA of cauliflower mosaic virus (Schmidt-Puchta et al. 1997; Dominguez et al. 1998; Hemmings-Mieszczak and Hohn 1999) and the late RNAs of adenovirus (Yueh and Schneider 1996). Ribosomal shunting is thought to require a low amount of active or phosphorylated eIF4E, as the unwinding activity of the eIF4F complex is not needed for the shunt. This could explain the relatively efficient translation of HSP70 mRNAs not only in heat-shocked cells, in which formation of eIF4F is largely reduced, but also in cells in which eIF4F formation is inhibited by other means (Joshi-Barve et al. 1992; Novoa and Carrasco 1999).

We extended our studies in rat hepatoma H35 cells, in which heat shock caused an increased phosphorylation of 4E-BP1 and increased levels of 'free' eIF4E, to assess the role of eIF4F formation in the inhibition of protein synthesis and preferential translation of HSP mRNAs after stress. Besides heat shock, arsenite treatment of cells was included, as it has been shown to increase eIF4E phosphorylation.

As mentioned above, heat shock causes eIF4E dephosphorylation, whereas arsenite can cause the opposite effect. If eIF4E phosphorylation plays an important role in the expression of the HSPs in mammalian cells, as suggested in the literature, one would expect differences in the HSP expression patterns following these different stress treatments. In Fig. 1A the expression of heat shock proteins was determined in recovered cells.

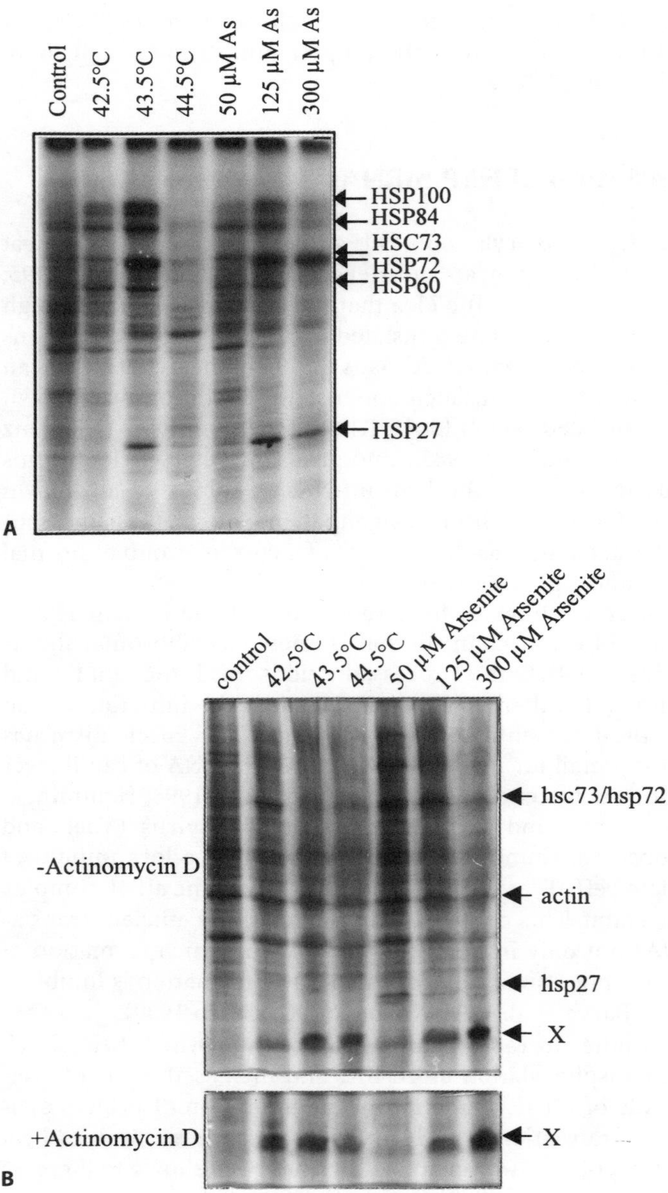

Fig. 1A,B. HSP expression in stressed cells. Samples of recovered cells as described in Table 1, containing 100,000 cpm of hot-TCA precipitable labelled protein were separated on 12.5% polyacrylamide gels (**A**). Samples were also taken from cells that were labelled immediately following the stress treatment for 1 h either in the absence or presence of 0.5 μg actinomycin D/ml (**B**). Loading larger samples interferes with the proper running of the gel, because the lanes will be overloaded with protein. Gels were fixed, treated with sodium salicylate, and used for fluorography. Induced HSPs are indicated on the *right*

Heat shock, or exposure to arsenite, resulted in very similar HSP expression in H35 cells (Fig. 1A). Five major inducible HSPs (27, 60, 72, 84, and 100) were seen in both cases, and synthesis of the constitutively expressed HSC73 was increased as well (Wiegant et al. 1994). The 44.5°C treatment of the cells had a very severe effect on general protein synthesis, as also reflected in the reduced ability to recover (Table 1). In Fig. 1, a 5-h recovery period was chosen to obtain a clear profile of the induced HSPs. We performed a similar analysis immediately after the stress treatment, to minimize the influence of de novo HSP mRNA transcription (Fig. 1B, upper part). The patterns of proteins synthesized were similar for heat-shocked and arsenite-treated cells in the first hour after the stress, HSP72/HSC73 being one of the few proteins that is already synthesized. Actinomycin D inhibited the induction of HSP72 and HSC73 even within the first hour (not shown), showing that major changes in HSP levels are transcriptionally regulated. The exception is an unknown protein of approximately 20 kDa (marked with an "X"), whose expression was strongly increased in both arsenite-treated and heat-shocked cells, and was not affected by actinomycin D (Fig. 1B, lower part). Apparently, the mRNA of this protein is selectively translated under both stress conditions, despite the opposite effects of the two treatments on eIF4E phosphorylation and eIF4F formation (Duncan and Hershey 1989; Duncan et al. 1995; Scheper et al. 1997; Vries et al. 1997; Wang and Proud 1997; Wang et al. 1998; see also Figs. 2 and 3 below).

Heat shock caused a decrease in eIF4E phosphorylation, especially at the higher temperatures (Fig. 2). In contrast, arsenite treatment of H35 cells induced an almost complete conversion of eIF4E into its phosphorylated form. In spite of this remarkable difference in eIF4E phosphorylation, both treatments induced 4E-BP1 phosphorylation. Phosphorylation of eIF4E and 4E-BP1 appears to be a separately regulated process; eIF4E dephosphorylation can occur while 4E-BP1 is phosphorylated in heat-shocked cells (Fig. 2). Similar observations were made in other cell types (Feigenblum and Schneider 1996; Scheper et al. 1997) Therefore, even though 4E-BP1 binding might have a negative effect on eIF4E phosphorylation (Waskiewicz et al. 1997; Wang et al. 1998), more 'free' eIF4E does not automatically lead to increased eIF4E phosphorylation. It has been suggested that phosphorylation of eIF4E by Mnk1 in vivo could be affected by binding of 4E-BP1 to eIF4E (Wang et al. 1998). For example, hydrogen peroxide or sorbitol treatment of 293 cells activated Mnk1 but did not lead to increased eIF4E phosphorylation. This discrepancy was thought to be due to decreased 4E-BP1 phosphorylation, leading to formation of the 4E-BP1/eIF4E complex and less efficient Mnk1-induced phosphorylation of eIF4E in the 4E-BP1/eIF4E complex (Wang et al. 1998). Our results with H35 cells show that eIF4E phosphorylation in stressed cells is not exclusively regulated through 4E-BP1 binding, since heat shock as well as arsenite induced 4E-BP1 phosphorylation, but only arsenite treatment resulted in increased eIF4E phosphorylation. The differences in eIF4E phosphorylation upon different stress treatments are more likely to be caused by differences in the effects on the Mnk1/eIF4G or eIF4E/eIF4G association than in the eIF4E/4E-BP1 asso-

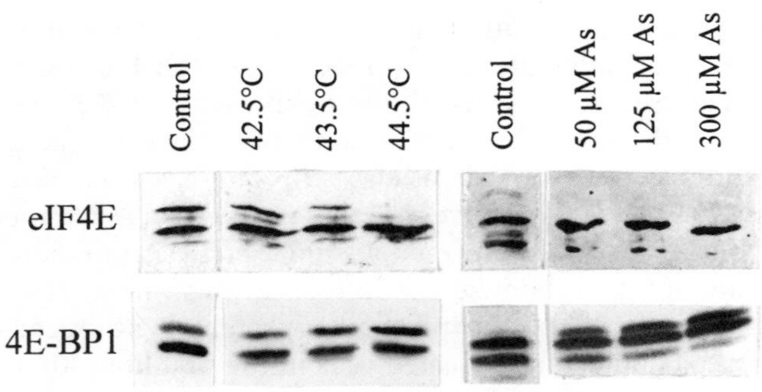

Fig. 2. eIF4E and 4E-BP1 phosphorylation in stressed cells. H35 cells were treated for 30 min at the indicated temperatures or incubated with arsenite as indicated for 1 h. Cells were harvested as described in the experimental procedures (under m^{7}GTP-Sepharose purification). *Lower panel* Extracts were analysed by SDS-PAGE and Western blotting to determine the 4E-BP1 phosphorylation state. Slower migrating forms are the more phosphorylated forms (Lin et al. 1994; Scheper et al. 1997). *Upper panel* eIF4E was purified by m^{7}GTP-Sepharose affinity and analysed by isoelectric focusing and Western blotting to determine the eIF4E phosphorylation state. The *upper band* represents the phosphorylated form of eIF4E

ciation. However, a detailed study on the effects of Mnk1 binding to eIF4G under a variety of stress conditions has not been reported yet.

The two recent papers on Mnk1 (Pyronnet et al. 1999; Waskiewicz et al. 1999) show contradicting results on the importance of Mnk1 binding to eIF4G in vivo. Overexpression of 4E-BP1 did not interfere with Mnk1 activity in transfected cells, indicating that eIF4E could be phosphorylated as part of the eIF4E/4E-BP1 complex (Waskiewicz et al. 1999). However, these results do not exclude recycling of eIF4E through the eIF4F complex, where phosphorylation could occur. The inhibition of Mnk1 in cells expressing a mutant of eIF4G that could not bind eIF4E (Pyronnet et al. 1999) seems to stress the importance of Mnk1 binding to eIF4G in vivo.

To study in more detail the importance of eIF4E binding to eIF4G for the phosphorylation of eIF4E, we compared eIF4F formation in heat-shocked and arsenite-treated cells. eIF4E and associated eIF4G and 4E-BP1 were purified by m^{7}GTP-Sepharose (Fig. 3). Both kinds of stress treatment resulted in increased phosphorylation of 4E-BP1 and, as a consequence, less 4E-BP1 was associated with eIF4E under these conditions (cf. Fig. 2 with Fig. 3).

Increasing temperatures resulted in a severe decrease in bound eIF4G. The loss of eIF4G could be partly explained by degradation of eIF4G, seen as a reduction of eIF4G in the extract. However, the effect of heat shock on eIF4G binding is much more extensive than can be explained by degradation alone. Apparently, less eIF4G was bound to eIF4E in heat-shocked cells. The cause for this dissociation remains to be elucidated. In clear contrast, arsenite treatment led to increased eIF4G association with eIF4E. The increase in eIF4E-bound eIF4G in arsenite-treated cells was not caused by an increased amount of eIF4G

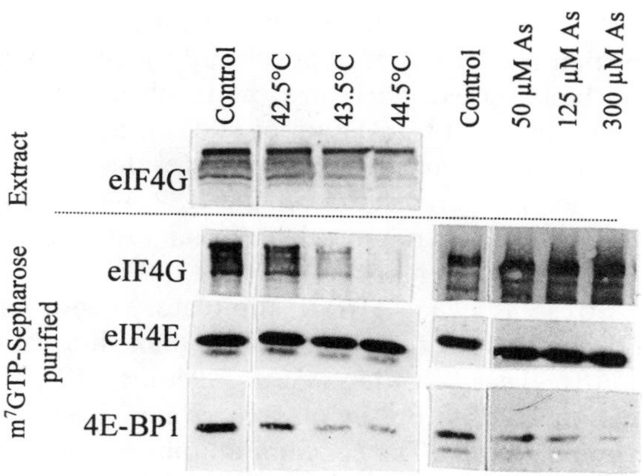

Fig. 3. Association of 4E-BP1 and eIF4G to eIF4E in stressed cells. H35 cells were treated for 30 min at the indicated temperatures or incubated with arsenite as indicated for 1 h. After these treatments, the cells were harvested and used for purification of eIF4E and associated proteins as described in experimental procedures. Samples purified by m^7GTP-Sepharose were separated on 12.5% polyacrylamide gels and analysed by Western blotting with antibodies against eIF4G (m^7GTP-Sepharose purified *upper panel*), eIF4E (m^7GTP-Sepharose purified *middle panel*), and 4E-BP1 (m^7GTP-Sepharose purified *lower panel*). As a control for the amounts of eIF4G in these cells, cell extracts prior to m^7GTP-Sepharose purification were analysed as well (*upper panel* "extract")

(not shown). The heat instability of eIF4G becomes more evident above 44.5 °C and is accompanied by a reduction in eIF4E levels as well (not shown). Apparently, eIF4G integrity is easily lost upon heat treatment. The instability of eIF4G has been known for years, and is reflected in the difficulty of purifying this initiation factor in an intact form.

The results for the heat-shocked cells could favour a model in which decreased activity of the cap-binding complex eIF4F plays an important role in the assumed preferential translation of the HSPs. Even though the availability of eIF4E is increased by release of the inhibitory binding proteins, eIF4F formation is diminished. Thus, a close association or interaction between Mnk1 and eIF4E will not be established. This will eventually result in decreased eIF4E phosphorylation, through dephosphorylation by phosphatases. It should be noted that the regulation of the activity of phosphatases, which have phosphorylated translation factors as substrates, has not been studied in any great detail yet. The results on eIF4E phosphorylation obtained from our work on arsenite treatment of cells strongly contradict the suggested role for eIF4F in the translation of stress proteins.

Arsenite treatment of H35 cells led to optimal conditions for the formation of an active eIF4F complex: (1) increased 4E-BP1 phosphorylation and dissociation from eIF4E; (2) increased eIF4E phosphorylation; (3) increased eIF4G association with eIF4E. All these phenomena are expected to enhance cap-dependent translation, which is in contrast to the general inhibition of protein synthesis (Fig. 1). Apparently, depending on cell type (Scheper et al. 1997) or the nature of the stress, opposite effects on various aspects of eIF4F formation can be observed. Therefore, the involvement of eIF4E, 4E-BP1, and eIF4G in the preferential translation of HSP mRNAs under stress conditions in mammalian cells seems debatable. Our results, shown in Fig. 1, failed to show such a preferential translation of the HSP mRNAs immediately after stress treatment of rat hepatoma cells, similar to the results obtained with HeLa cells (Duncan and Hershey 1989).

Stress, as well as growth stimulatory conditions, have been shown to cause a diverse array of effects on eIF4E phosphorylation and 4E-BP1 phosphorylation (reviewed in Kleijn et al. 1998a). eIF4E phosphorylation and eIF4F formation seem to be the only correlated phenomena in stressed as well as in stimulated cells. The concomitant phosphorylation of eIF4E and eIF4E/eIF4G binding indicates either that eIF4E phosphorylation occurs mainly in the eIF4F complex, as supported by in vitro data (Tuazon et al. 1990), and by the finding that Mnk1 binds to eIF4G (Pyronnet et al. 1999; Waskiewicz et al. 1999), or that eIF4E phosphorylation drives the formation of eIF4F. From our results with arsenite-treated cells, it is very clear that enhanced eIF4F formation does not necessarily lead to increased protein synthesis, or that active protein synthesis is required for eIF4F formation.

The mechanism by which the HSP mRNAs can be translated under conditions where general protein synthesis is reduced remains unclear. The 5′ UTRs of (some of) the HSP mRNAs are thought to confer the ability to be translated with a low requirement for active eIF4E or eIF4F. However, the known mechanisms by which this reduced activity of eIF4E can be achieved, i.e., increased binding to 4E-BP1, dephosphorylation, or reduced binding to eIF4G, are not a prerequisite for HSP translation per se. Perhaps, another, still undiscovered mechanism may lead to reduced eIF4E activity under all stress conditions. Hypothetically, this mechanism involves a reduced interaction of one of the components of the eIF4F complex with HSPs, as a reduced level of chaperone protein seems to be the major way in which a cell senses stress. The observed stress-dependent changes in eIF4E phosphorylation, and binding of eIF4E to either 4E-BP1 or eIF4G, might enhance or reduce the severity of the inhibition of protein synthesis or the expression of the HSPs.

3 Experimental Procedures

Cells and Culture Conditions. Rat hepatoma reuber H35 cells were grown to a confluency of 80–90% in Leibowitz (L15) medium (Flow/ICN Laboratories) containing potassium penicillin G (100 units/ml), streptomycin sulfate (100 μg/ml), and 10% fetal calf serum (Gibco BRL).

Heat Treatment. Heat shocks were applied by submerging the culture flasks in a heated water bath which provides a temperature stable within 0.1 °C (±SE).

Under these conditions temperature equilibration of the cells took about 0.5 min.

Arsenite Treatment. Cells were treated with sodium arsenite by adding to the medium the amounts indicated in the figure legends; the cells were incubated at 37 °C for 1 h.

Incorporation Studies. Cell labelling was carried out with 7 μCi of [^{35}S]-methionine/cysteine (Promix, Amersham) in 2 ml of HEPES-buffered DMEM lacking methionine. After harvesting of the cells, 10% of the sample was used to determine [^{35}S]-methionine/cysteine incorporation by hot-TCA precipitation. For analysis of the synthesis of heat shock proteins, 100,000 cpm of incorporated radioactivity, as determined by hot-TCA precipitation, was separated on 12.5% polyacrylamide gels, which were subsequently dried and used for autoradiography.

Isoelectric Focusing and eIF4E Phosphorylation. Cells were harvested in 75 μl of isoelectric focusing buffer (ISB) (9.5 M urea (Gibco BRL), 12 mM 3-[(3-cholamidopropyl)-dimethylammonio]-1-propanesulfonic acid (CHAPS) (Boehringer-Mannheim), 0.75% biolytes 3/10, 2.25% biolytes 4/6 (Biorad), and 700 mM ß-mercaptoethanol). One third of each sample was subjected to denaturing one-dimensional-isoelectric focusing as described (Kleijn et al. 1995). After blotting onto PVDF (polyvinylidene fluoride) membranes, phosphorylated and unphosphorylated forms of eIF4E were visualized with a polyclonal antibody and an alkaline phosphatase-conjugated secondary antibody.

m^{7}GTP-Sepharose Purification of eIF4E and Associated Proteins. Purification of eIF4E and associated proteins was performed essentially as described previously (Kleijn et al. 1995). Eluted eIF4E and associated proteins were analyzed by SDS-PAGE or isoelectric focusing and immunoblotting. After blotting of the SDS-polyacrylamide gel, the PVDF filter was cut into three pieces, and the separate parts were probed with antibodies against eIF4E, 4E-BP1, or eIF4G (a gift from Dr. R. Rhoads).

eIF2B Activity. eIF2·[^{3}H]GDP complexes were made as described (Mehta et al. 1983). Briefly, 1 pmol eIF2 was incubated with 0.2 μCi of [^{3}H]GDP (approx. 15 pmol, 30,000 dpm/pmol) in 20 mM Tris-HCl pH 7.6, 120 mM KCl, 1% BSA, and 1 mM DTT. After incubation at 30 °C for 15 min, 5 mM $MgCl_2$, 1 mM GTP, and cell extract (prepared as described below) was added.

Cells were harvested in 20 mM Tris-HCl pH 7.6, 100 mM KCl, 1% Triton X-100, 0.2 mM EDTA, 50 mM β-glycerophosphate, 1 mM sodium molybdate, 10% glycerol, 4 μg leupeptin/ml, 0.2 mM benzamidine, 0.2 mM sodium vanadate, and 7 mM β-mercaptoethanol. Cell extracts, approximately 10 μg total protein, were added to eIF2·[^{3}H]GDP complexes and incubated for 15 min at 30 °C. The GDP-GTP exchange reaction was stopped by adding 1 ml of a cold wash buffer (50 mM Tris-HCl pH 7.6, 5 mM $MgCl_2$, 100 mM KCl, and 7 mM β-mercaptoethanol). The mixture was filtered through nitrocellulose filters and washed three times with the same buffer. The activity of eIF2B was determined by quantification of the amount of eIF2·[^{3}H]GDP retained on the filter.

Acknowledgements. We thank Dr. R. Rhoads for his friendly gift of eIF4G antibody, and Hans van Aken and Marcelle Kasperaitis for technical assistance. This work was supported by the HomInt Organisation.

References

Barber GN, Thompson S, Lee TG, Strom T, Jagus R, Darveau A, Katze MG (1994) The 58-kilodalton inhibitor of the interferon-induced double-stranded RNA-activated protein kinase is a tetratricopeptide repeat protein with oncogenic properties. Proc Natl Acad Sci USA 91:4278–4282

Berlanga JJ, Herrero S, De Haro C (1998) Characterization of the hemin-sensitive eukaryotic initiation factor 2alpha kinase from mouse nonerythroid cells. J Biol Chem 273:32340–32346

Berlanga JJ, Santoyo J, De Haro C (1999) Characterization of a mammalian homolog of the GCN2 eukaryotic initiation factor 2alpha kinase. Eur J Biochem 265:754–762

Brostrom CO, Prostko CR, Kaufman RJ, Brostrom MA (1996) Inhibition of translational initiation by activators of the glucose-regulated stress protein and heat shock protein stress response systems. Role of the interferon-inducible double-stranded RNA-activated eukaryotic initiation factor 2alpha kinase. J Biol Chem 271:24995–25002

Chen JJ, London IM (1995) Regulation of protein synthesis by heme-regulated eIF-2 alpha kinase. Trends Biochem Sci 20:105–108

Chen JJ, Crosby JS, London IM (1994) Regulation of heme-regulated eIF-2 alpha kinase and its expression in erythroid cells. Biochimie 76:761–769

Cigan AM, Bushman JL, Boal TR, Hinnebusch AG (1993) A protein complex of translational regulators of GCN4 mRNA is the guanine nucleotide-exchange factor for translation initiation factor 2 in yeast. Proc Natl Acad Sci USA 90:5350–5354

Clemens MJ (1997) PKR – a protein kinase regulated by double-stranded RNA. Int J Biochem Cell Biol 29:945–949

Clemens MJ, Laing KG, Jeffrey IW, Schofield A, Sharp TV, Elia A, Matys V, James MC, Tilleray VJ (1994) Regulation of the interferon-inducible eIF-2 alpha protein kinase by small RNAs. Biochimie 76:770–778

Davies MV, Elroy-Stein O, Jagus R, Moss B, Kaufman RJ (1992) The vaccinia virus K3L gene product potentiates translation by inhibiting double-stranded-RNA-activated protein kinase and phosphorylation of the alpha subunit of eukaryotic initiation factor 2. J Virol 66: 1943–1950

Davies MV, Chang HW, Jacobs BL, Kaufman RJ (1993) The E3L and K3L vaccinia virus gene products stimulate translation through inhibition of the double-stranded RNA-dependent protein kinase by different mechanisms. J Virol 67:1688–1692

Dholakia JN, Mueser TC, Woodley CL, Parkhurst LJ, Wahba AJ (1986) The association of NADPH with the guanine nucleotide exchange factor from rabbit reticulocytes: a role of pyridine dinucleotides in eukaryotic polypeptide chain initiation. Proc Natl Acad Sci USA 83:6746–6750

Dholakia JN, Francis BR, Haley BE, Wahba AJ (1989) Photoaffinity labeling of the rabbit reticulocyte guanine nucleotide exchange factor and eukaryotic initiation factor 2 with 8-azidopurine nucleotides. Identification of GTP- and ATP-binding domains. J Biol Chem 264:20638–20642

Dominguez DI, Ryabova LA, Pooggin MM, Schmidt-Puchta W, Futterer J, Hohn T (1998) Ribosome shunting in cauliflower mosaic virus. Identification of an essential and sufficient structural element. J Biol Chem 273:3669–3678

Duncan RF (1996) Translational control during heat shock. In: Hershey JWB, Mathews MB, Sonenberg N (eds) Translational control. Cold Spring Harbor Laboratory Press, Cold Spring Harbor, pp 271–294

Duncan RF, Hershey JW (1987) Translational repression by chemical inducers of the stress response occurs by different pathways. Arch Biochem Biophys 256:651–661

Duncan RF, Hershey JW (1989) Protein synthesis and protein phosphorylation during heat stress, recovery, and adaptation. J Cell Biol 109:1467–1481

Duncan RF, Cavener DR, Qu S (1995) Heat shock effects on phosphorylation of protein synthesis initiation factor proteins eIF-4E and eIF-2 alpha in *Drosophila*. Biochemistry 34:2985–2997

Feigenblum D, Schneider RJ (1996) Cap-binding protein (eukaryotic initiation factor 4E) and 4E-inactivating protein BP-1 independently regulate cap-dependent translation. Mol Cell Biol 16:5450–5457

Flynn A, Proud CG (1995) Serine 209, not serine 53, is the major site of phosphorylation in initiation factor eIF-4E in serum-treated Chinese hamster ovary cells. J Biol Chem 270: 21684–21688

Flynn A, Vries RG, Proud CG (1997) Signalling pathways which regulate eIF4E. Biochem Soc Trans 25:192S

Fraser C, Morley S (1997) Studies on the phosphorylation of eIF4E in *Xenopus* (XIK-2) kidney cells. Biochem Soc Trans 25:190S

Fukunaga R, Hunter T (1997) MNK1, a new MAP kinase-activated protein kinase, isolated by a novel expression screening method for identifying protein kinase substrates. EMBO J 16:1921–1933

Gale M Jr, Blakely CM, Hopkins DA, Melville MW, Wambach M, Romano PR, Katze MG (1998) Regulation of interferon-induced protein kinase PKR: modulation of P58IPK inhibitory function by a novel protein, P52rIPK. Mol Cell Biol 18:859–871

Haghighat A, Sonenberg N (1997) eIF4G dramatically enhances the binding of eIF4E to the mRNA 5′-cap structure. J Biol Chem 272:21677–21680

Harding HP, Zhang Y, Ron D (1999) Protein translation and folding are coupled by an endoplasmic-reticulum-resident kinase. Nature 397:271–274

He B, Gross M, Roizman B (1997) The gamma(1)34.5 protein of herpes simplex virus 1 complexes with protein phosphatase 1alpha to dephosphorylate the alpha subunit of the eukaryotic translation initiation factor 2 and preclude the shutoff of protein synthesis by double-stranded RNA-activated protein kinase. Proc Natl Acad Sci USA 94:843–848

Hemmings-Mieszczak M, Hohn T (1999) A stable hairpin preceded by a short open reading frame promotes nonlinear ribosome migration on a synthetic mRNA leader. RNA 5:1149–1157

Hiremath LS, Hiremath ST, Rychlik W, Joshi S, Domier LL, Rhoads RE (1989) In vitro synthesis, phosphorylation, and localization on 48 S initiation complexes of human protein synthesis initiation factor 4E. J Biol Chem 264:1132–1138 (Erratum in J Biol Chem 264:21431)

Ito T, Yang M, May WS (1999) RAX, a cellular activator for double-stranded RNA-dependent protein kinase during stress signaling. J Biol Chem 274:15427–15432

Joshi CP, Nguyen HT (1995) 5′ untranslated leader sequences of eukaryotic mRNAs encoding heat shock induced proteins. Nucleic Acids Res 23:541–549

Joshi B, Cai AL, Keiper BD, Minich WB, Mendez R, Beach CM, Stepinski J, Stolarski R, Darzynkiewicz E, Rhoads RE (1995) Phosphorylation of eukaryotic protein synthesis initiation factor 4E at Ser-209. J Biol Chem 270:14597–14603

Joshi-Barve S, De Benedetti A, Rhoads RE (1992) Preferential translation of heat shock mRNAs in HeLa cells deficient in protein synthesis initiation factors eIF-4E and eIF-4 gamma. J Biol Chem 267:21038–21043

Kimball SR, Fabian JR, Pavitt GD, Hinnebusch AG, Jefferson LS (1998) Regulation of guanine nucleotide exchange through phosphorylation of eukaryotic initiation factor eIF2alpha. Role of the alpha- and delta-subunits of eIF2b. J Biol Chem 273:12841–12845

Kleijn M, Voorma HO, Thomas AAM (1995) Phosphorylation of eIF-4E and initiation of protein synthesis in P19 embryonal carcinoma cells. J Cell Biochem 59:443–452

Kleijn M, Scheper GC, Voorma HO, Thomas AAM (1998a) Regulation of translation initiation factors by signal transduction. Eur J Biochem 253:531–544

Kleijn M, Welsh GI, Scheper GC, Voorma HO, Proud CG, Thomas AAM (1998b) Nerve and epidermal growth factor induce protein synthesis and eIF2B activation in PC12 cells. J Biol Chem 273:5536–5541

Klemenz R, Hultmark D, Gehring WJ (1985) Selective translation of heat shock mRNA in Drosophila melanogaster depends on sequence information in the leader. EMBO J 4:2053–2060

Lamphear BJ, Kirchweger R, Skern T, Rhoads RE (1995) Mapping of functional domains in eukaryotic protein synthesis initiation factor 4G (eIF4G) with picornaviral proteases. Implications for cap-dependent and cap-independent translational initiation. J Biol Chem 270:21975–21983

Lamphear BJ, Panniers R (1990) Cap binding protein complex that restores protein synthesis in heat-shocked Ehrlich cell lysates contains highly phosphorylated eIF-4E. J Biol Chem 265:5333–5336

Lin TA, Kong X, Haystead TA, Pause A, Belsham G, Sonenberg N, Lawrence JC Jr (1994) PHAS-I as a link between mitogen-activated protein kinase and translation initiation. Science 266:653–656 (Comments)

Lindquist S, Petersen R (1990) Selective translation and degradation of heat-shock messenger RNAs in *Drosophila*. Enzyme 44:147–166

Mader S, Lee H, Pause A, Sonenberg N (1995) The translation initiation factor eIF-4E binds to a common motif shared by the translation factor eIF-4 gamma and the translational repressors 4E-binding proteins. Mol Cell Biol 15:4990–4997

Matts RL, Xu Z, Pal JK, Chen JJ (1992) Interactions of the heme-regulated eIF-2 alpha kinase with heat shock proteins in rabbit reticulocyte lysates. J Biol Chem 267:18160–18167

Matts RL, Hurst R, Xu Z (1993) Denatured proteins inhibit translation in hemin-supplemented rabbit reticulocyte lysate by inducing the activation of the heme-regulated eIF-2 alpha kinase. Biochemistry 32:7323–7328

McGarry TJ, Lindquist S (1985) The preferential translation of Drosophila hsp70 mRNA requires sequences in the untranslated leader. Cell 42:903–911

Mehta HB, Woodley CL, Wahba AJ (1983) Protein synthesis in brine shrimp embryos and rabbit reticulocytes. The effect of Mg^{2+} on binary (eukaryotic initiation factor 2 X GDP) and ternary (eukaryotic initiation factor 2 X GTP X met-tRNAf) complex formation. J Biol Chem 258:3438–3441

Melville MW, Hansen WJ, Freeman BC, Welch WJ, Katze MG (1997) The molecular chaperone hsp40 regulates the activity of P58IPK, the cellular inhibitor of PKR. Proc Natl Acad Sci USA 94:97–102

Meurs EF, Galabru J, Barber GN, Katze MG, Hovanessian AG (1993) Tumor suppressor function of the interferon-induced double-stranded RNA- activated protein kinase. Proc Natl Acad Sci USA 90:232–236

Minich WB, Balasta ML, Goss DJ, Rhoads RE (1994) Chromatographic resolution of in vivo phosphorylated and nonphosphorylated eukaryotic translation initiation factor eIF-4E: increased cap affinity of the phosphorylated form. Proc Natl Acad Sci USA 91:7668–7672

Morino S, Imataka H, Svitkin YV, Pestova TV, Sonenberg N (2000) Eukaryotic translation initiation factor 4E (eIF4E) binding site and the middle one-third of eIF4GI constitute the core domain for cap-dependent translation, and the C-terminal one-third functions as a modulatory region. Mol Cell Biol 20:468–477

Morley SJ (1997) Signalling through either the p38 or ERK mitogen-activated protein (MAP) kinase pathway is obligatory for phorbol ester and T cell receptor complex (TCR-CD3)-stimulated phosphorylation of initiation factor (eIF) 4E in Jurkat T cells. FEBS Lett 418:327–332

Novoa I, Carrasco L (1999) Cleavage of eukaryotic translation initiation factor 4G by exogenously added hybrid proteins containing poliovirus 2Apro in HeLa cells: effects on gene expression. Mol Cell Biol 19:2445–2454

Pain VM (1996) Initiation of protein synthesis in eukaryotic cells. Eur J Biochem 236:747–771
Panniers R, Henshaw EC (1983) A GDP/GTP exchange factor essential for eukaryotic initiation factor 2 cycling in Ehrlich ascites tumor cells and its regulation by eukaryotic initiation factor 2 phosphorylation. J Biol Chem 258:7928–7934
Patel RC, Sen GC (1998) PACT, a protein activator of the interferon-induced protein kinase, PKR. EMBO J 17:4379–4390
Pause A, Belsham GJ, Gingras A-C, Donzé O, Lin T-A, Lawrence JC Jr, Sonenberg N (1994) Insulin-dependent stimulation of protein synthesis by phosphorylation of a regulator of 5′-cap function. Nature 371:762–767
Price N, Proud C (1994) The guanine nucleotide-exchange factor, eIF-2B. Biochimie 76:748–760
Price NT, Welsh GI, Proud CG (1991) Phosphorylation of only serine-51 in protein synthesis initiation factor-2 is associated with inhibition of peptide-chain initiation in reticulocyte lysates. Biochem Biophys Res Commun 176:993–999
Proud CG (1992) Protein phosphorylation in translational control. Curr Topics Cell Regul 32:243–369
Ptushkina M, von der Haar T, Karim MM, Hughes JM, McCarthy JE (1999) Repressor binding to a dorsal regulatory site traps human eIF4E in a high cap-affinity state. EMBO J 18:4068–4075
Pyronnet S, Imataka H, Gingras AC, Fukunaga R, Hunter T, Sonenberg N (1999) Human eukaryotic translation initiation factor 4G (eIF4G) recruits mnk1 to phosphorylate eIF4E. EMBO J 18:270–279
Rhoads RE, Lamphear BJ (1995) Cap-independent translation of heat shock messenger RNAs. In: Sarnow P (ed) Cap-independent translation. Springer, Berlin Heidelberg New York, pp 131–153
Scheper GC, Mulder J, Kleijn M, Voorma HO, Thomas AAM, van Wijk R (1997) Inactivation of eIF2B and phosphorylation of PHAS-I in heat-shocked rat hepatoma cells. J Biol Chem 272:26850–26856
Scheper GC, Thomas AA, van Wijk R (1998) Inactivation of eukaryotic initiation factor 2B in vitro by heat shock. Biochem J 334:463–467
Schmidt-Puchta W, Dominguez D, Lewetag D, Hohn T (1997) Plant ribosome shunting in vitro. Nucleic Acids Res 25:2854–2860
Sharp TV, Moonan F, Romashko A, Joshi B, Barber GN, Jagus R (1998) The vaccinia virus E3L gene product interacts with both the regulatory and the substrate binding regions of PKR: implications for PKR autoregulation. Virology 250:302–315
Shi Y, Vattem KM, Sood R, An J, Liang J, Stramm L, Wek RC (1998) Identification and characterization of pancreatic eukaryotic initiation factor 2 alpha-subunit kinase, PEK, involved in translational control. Mol Cell Biol 18:7499–7509
Shi Y, An J, Liang J, Hayes SE, Sandusky GE, Stramm LE, Yang NN (1999) Characterization of a mutant pancreatic eIF-2alpha kinase, PEK, and co-localization with somatostatin in islet delta cells. J Biol Chem 274:5723–5730
Sood R, Porter AC, Olsen D, Cavener DR, Wek RC (2000) A mammalian homologue of GCN2 protein kinase important for translational control by phosphorylation of eukaryotic initiation factor-2alpha. Genetics 154:787–801
Srivastava SP, Davies MV, Kaufman RJ (1995) Calcium depletion from the endoplasmic reticulum activates the double- stranded RNA-dependent protein kinase (PKR) to inhibit protein synthesis. J Biol Chem 270:16619–16624
Tarun SZ Jr, Sachs AB (1996) Association of the yeast poly(A) tail binding protein with translation initiation factor eIF-4G. EMBO J 15:7168–7177
Thulasiraman V, Xu Z, Uma S, Gu Y, Chen JJ, Matts RL (1998) Evidence that Hsc70 negatively modulates the activation of the heme- regulated eIF-2alpha kinase in rabbit reticulocyte lysate. Eur J Biochem 255:552–562
Tuazon PT, Morley SJ, Dever TE, Merrick WC, Rhoads RE, Traugh JA (1990) Association of initiation factor eIF-4E in a cap binding protein complex (eIF-4F) is critical for and enhances phosphorylation by protein kinase C. J Biol Chem 265:10617–10621

Uma S, Hartson SD, Chen JJ, Matts RL (1997) Hsp90 is obligatory for the heme-regulated eIF-2alpha kinase to acquire and maintain an activable conformation. J Biol Chem 272:11648–11656 (Erratum in J Biol Chem 272:16068)

Uma S, Thulasiraman V, Matts RL (1999) Dual role for Hsc70 in the biogenesis and regulation of the heme-regulated kinase of the alpha subunit of eukaryotic translation initiation factor 2. Mol Cell Biol 19:5861–5871

Vries RG, Flynn A, Patel JC, Wang X, Denton RM, Proud CG (1997) Heat shock increases the association of binding protein-1 with initiation factor 4E. J Biol Chem 272:32779–32784

Wang X, Proud CG (1997) p70 S6 kinase is activated by sodium arsenite in adult rat cardiomyocytes: roles for phosphatidylinositol 3-kinase and p38 Map kinase. Biochem Biophys Res Commun 238:207–212

Wang X, Flynn A, Waskiewicz AJ, Webb BL, Vries RG, Baines IA, Cooper JA, Proud CG (1998) The phosphorylation of eukaryotic initiation factor eIF4E in response to phorbol esters, cell stresses, and cytokines is mediated by distinct MAP kinase pathways. J Biol Chem 273:9373–9377

Waskiewicz AJ, Flynn A, Proud CG, Cooper JA (1997) Mitogen-activated protein kinases activate the serine/threonine kinases Mnk1 and Mnk2. EMBO J 16:1909–1920

Waskiewicz AJ, Johnson JC, Penn B, Mahalingam M, Kimball SR, Cooper JA (1999) Phosphorylation of the cap-binding protein eukaryotic translation initiation factor 4E by protein kinase Mnk1 in vivo. Mol Cell Biol 19:1871–1880

Welsh GI, Proud CG (1993) Glycogen synthase kinase-3 is rapidly inactivated in response to insulin and phosphorylates eukaryotic initiation factor eIF-2B. Biochem J 294:625–629

Welsh GI, Loughlin AJ, Foulstone EJ, Price NT, Proud CG (1997a) Regulation of initiation factor eIF-2B by GSK-3 regulated phosphorylation. Biochem Soc Trans 25:191 S

Welsh GI, Stokes CM, Wang X, Sakaue H, Ogawa W, Kasuga M, Proud CG (1997b) Activation of translation initiation factor eIF2B by insulin requires phosphatidyl inositol 3-kinase. FEBS Lett 410:418–422

Wiegant FA, Souren JE, van Rijn J, van Wijk R (1994) Stressor-specific induction of heat shock proteins in rat hepatoma cells. Toxicology 94:143–159

Wu C (1995) Heat shock transcription factors: structure and regulation. Annu Rev Cell Dev Biol 11:441–469

Xu Z, Pal JK, Thulasiraman V, Hahn HP, Chen JJ, Matts RL (1997) The role of the 90-kDa heat-shock protein and its associated cohorts in stabilizing the heme-regulated eIF-2alpha kinase in reticulocyte lysates during heat stress. Eur J Biochem 246:461–470

Yueh A, Schneider RJ (1996) Selective translation initiation by ribosome jumping in adenovirus-infected and heat-shocked cells. Genes Dev 10:1557–1567

Yueh A, Schneider RJ (2000) Translation by ribosome shunting on adenovirus and hsp70 mRNAs facilitated by complementarity to 18 S rRNA. Genes Dev 14:414–421

Initiation Factor eIF2α Phosphorylation in Stress Responses and Apoptosis

Michael J. Clemens[1]

1 Introduction

Protein synthesis in eukaryotes is a complex process which can be regulated at many points in the pathway. In recent years it has become clear that both the overall rate of translation and the relative rates of synthesis of individual proteins can be controlled post-transcriptionally through changes in the activities or levels of a small number of key components, and that such regulation usually takes place at the level of polypeptide chain initiation. A large body of evidence indicates that the essential polypeptide chain initiation factor eIF2 is a frequent target for regulation, and that its activity is often rate-limiting for protein synthesis. The phosphorylation of the smallest (α) subunit of eIF2 is a widely used mechanism of translational control in many organisms, and there are numerous physiologically important situations where eIF2α kinases are activated or inhibited. This chapter provides a review of our knowledge concerning the mechanisms by which eIF2 is controlled by reversible protein phosphorylation.

A great variety of intracellular and extracellular influences can alter rates of protein synthesis. One important class of such influences can be grouped under the general heading of cellular stresses. In the broadest sense, these range from normal physiological changes such as variations in nutrient availability or the actions of growth factors and cytokines to pathological conditions such as virus infection, hyperthermia or the presence of toxic compounds. In some cases, these conditions can result in cell death by apoptosis. Modulation of eIF2α phosphorylation is often involved in these situations. The present state of knowledge of how this is brought about and what the physiological consequences may be for the activities of the cell, including its survival or death, are reviewed here.

Limitations of space preclude a detailed account of the mechanism of polypeptide chain initiation and the reader is referred to recent comprehensive reviews (Pain 1996; Kozak 1999; Preiss and Hentze 1999) for such information. Briefly, initiation factor eIF2 catalyses the binding of the initiator

[1] Department of Biochemistry and Immunology, St George's Hospital Medical School, University of London, Cranmer Terrace, London SW17 0RE, UK

Progress in Molecular and Subcellular Biology, Vol. 27
Robert E. Rhoads (Ed.)

$Met\text{-}tRNA_f$ to the 40 S ribosomal subunit, in a process which requires the formation of a ternary complex between eIF2, $Met\text{-}tRNA_f$ and a molecule of GTP (Fig. 1A). This complex associates with mRNA by a process requiring several other initiation factors (Pain 1996) and locates the initiating AUG codon, usually as a result of ribosomal scanning from the 5′ end of the mRNA. Exceptionally, the AUG may be recognised by a process of internal ribosome binding involving an internal ribosome entry site (IRES) (Jackson et al. 1995). Following the location of the correct AUG, two further events must occur. The GTP that was associated with the eIF2 molecule is hydrolysed to GDP and phosphate, concomitant with the dissociation of the initiation factors from the ribosome. GTP hydrolysis requires the involvement of the initiation factor eIF5. The other event is the binding of the 60 S ribosomal subunit to the initiation complex in order to form the complete 80 S ribosomal initiation complex. The GDP generated by hydrolysis of GTP remains associated with the eIF2 and has to be exchanged for another molecule of GTP before the eIF2 can be utilised for a subsequent round of protein synthesis. This exchange is catalysed by the guanine nucleotide exchange factor eIF2B (Webb and Proud 1997).

eIF2 contains three non-identical subunits, α, β and γ. These have been cloned from a number of eukaryotic species and their sequences show several conserved features (Kimball 1999). Regulation of eIF2 activity is frequently a consequence of changes in the phosphorylation state of the α subunit, although other mechanisms can also control this factor. Identification of Ser^{51} of eIF2α as the site that is phosphorylated has enabled site-directed mutagenesis studies to establish the functional significance of this modification. Phosphorylation at Ser^{51} leads to an increased affinity of the initiation factor for eIF2B, thus increasing the proportion of the latter that is trapped as an inactive complex with phosphorylated eIF2 and GDP (Rowlands et al. 1988b; Fig. 1B). Since eIF2B is usually present in vivo in lower abundance than eIF2 (Rowlands et al. 1988a, b; Kimball et al. 1994; Oldfield et al. 1994) the reduction in free eIF2B inhibits the overall rate of guanine nucleotide exchange on the remaining unphosphorylated eIF2.

In keeping with the fact that Ser^{51} of eIF2α, or the equivalent residue in other species, is the amino acid that is phosphorylated by a variety of related protein

Fig. 1A,B. Role of eIF2 in initiation of protein synthesis and the mechanism of inhibition by phosphorylation of eIF2α. **A** Under conditions of active protein synthesis eIF2 binds one molecule of GTP (to the γ subunit) and one molecule of initiator $Met\text{-}tRNA_f$ (to the β and/or γ subunits) and catalyses the binding of the $Met\text{-}tRNA_f$ to the 40 S ribosomal initiation complex (Kimball 1999). The GDP that is generated by the hydrolysis of the GTP during this process is subsequently exchanged for another molecule of GTP by the action of the guanine nucleotide exchange factor eIF2B (Webb and Proud 1997). This must take place before the eIF2 can participate in a further round of initiation. The eIF2 contacts eIF2B via a surface contributed by the α, β and δ subunits of the latter protein (Pavitt et al. 1998). **B** Following its phosphorylation by eIF2α-specific protein kinases eIF2 can take part in one round of initiation but, following GTP hydrolysis, then forms a stable complex with eIF2B. This sequesters the latter in a form which is unavailable to catalyse further guanine nucleotide exchange

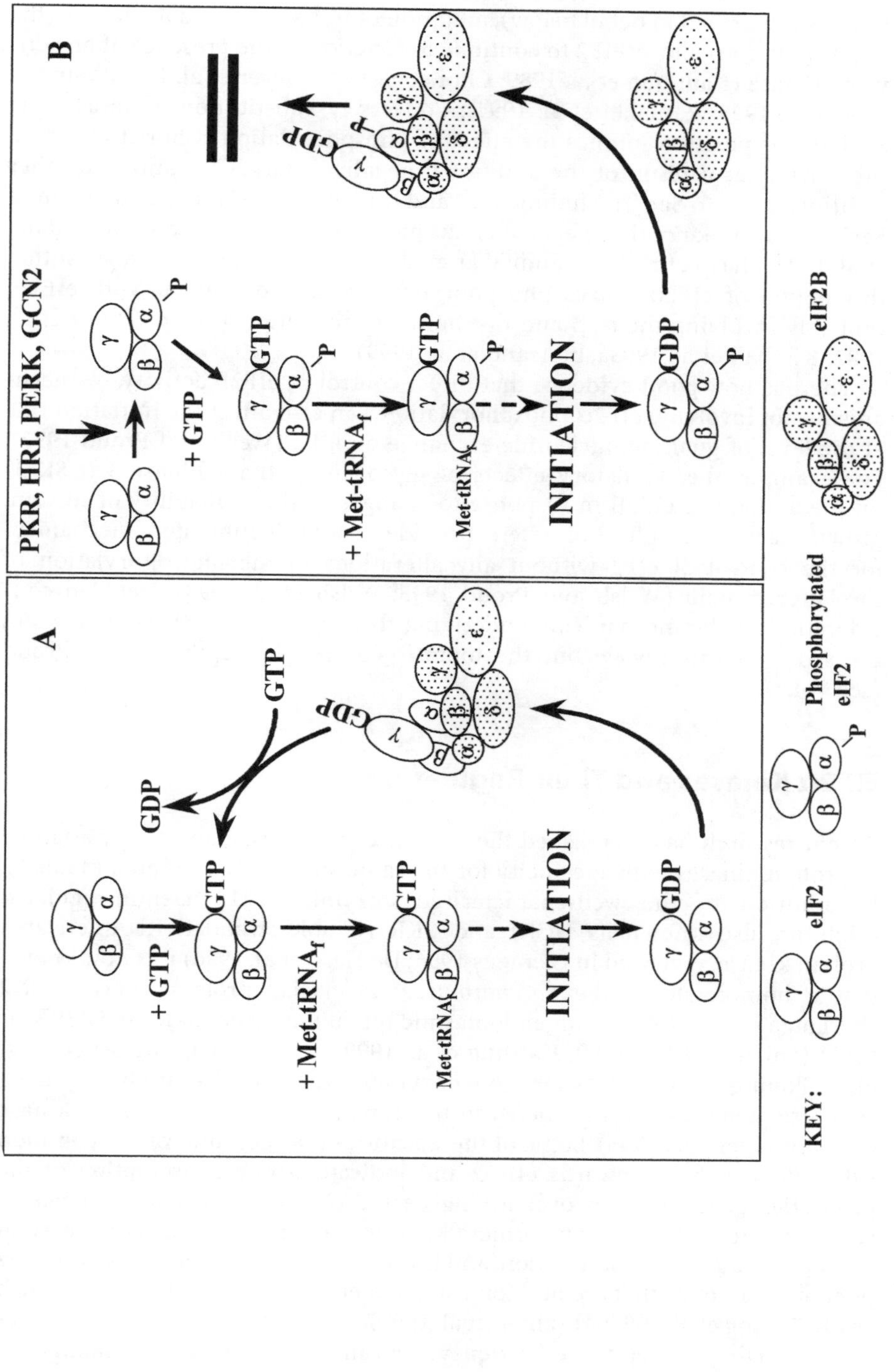
A
+ GTP
GTP
GDP
+ Met-tRNAf
Met-tRNAf
INITIATION
GDP
B
PKR, HRI, PERK, GCN2
+ GTP
GTP
+ Met-tRNAf
Met-tRNAf
INITIATION
GDP
KEY:
eIF2
Phosphorylated eIF2
eIF2B

kinases (described in detail below), mutation of this amino acid to alanine (the S51 A mutation) allows eIF2 to continue to function in the presence of an activated kinase (Kaufman et al. 1989; Choi et al. 1992; Dever et al. 1992; Murtha-Riel et al. 1993; Ramaiah et al. 1994). Conversely, substitution of an aspartic acid at this position mimics the effect of phosphorylation (Choi et al. 1992), presumably as a result of the additional negative charge. Mutations at other positions close to Ser^{51}, including Ser^{48} and the sequence GYID downstream of Ser^{51}, can also reduce the effects of eIF2α phosphorylation (Vazquez de Aldana et al. 1993; Sharp et al. 1997; Sudhakar et al. 1999). These findings suggest that this region of eIF2α makes phosphorylation-sensitive contacts with eIF2B, probably involving the α, β and δ subunits of the latter protein (Pavitt et al. 1997; Kimball et al. 1998a, b; Asano et al. 1999).

There is now good evidence that direct control of eIF2B activity, by mechanisms not involving eIF2α phosphorylation, can also affect the initiation rate at the level of guanine nucleotide exchange on eIF2 (Webb and Proud 1997). For example, the regulatory effects of glycogen synthase kinase-3 (GSK-3) on the activity of eIF2B in response to changes in the availability of insulin, growth factors or phorbol esters provide a well-documented mechanism for the control of eIF2 without any alterations in the phosphorylation of the latter protein (Welsh and Proud 1993; Welsh et al. 1998). Detailed consideration of the mechanisms underlying this regulatory pathway is beyond the scope of this review but the topic is covered in depth by C.G. Proud (see Vol. I).

2
EIF2α Kinases and Their Regulation

Recent research has established the existence in mammalian cells of a family of protein kinases with specificity for the same site on eIF2α (Ser^{51}) (Table 1). In addition to the well-characterised enzymes HRI (haemin-regulated inhibitor, also known as HCR) and PKR (double-stranded RNA-activated protein kinase; reviewed in Clemens 1996; De Haro et al. 1996) it is now known that a homologue of the *Saccharomyces cerevisiae* protein kinase GCN2 (Berlanga et al. 1999) and an endoplasmic reticulum protein kinase (PERK or PEK) (Shi et al. 1998, 1999; Harding et al. 1999) also occur in higher eukaryotes. Cloning of the cDNAs for these enzymes has revealed a number of interesting sequence similarities between them, particularly in the protein kinase domains (Clemens 1996). Some of these features may account for the common substrate specificity towards eIF2α and indicate strong conservation of the properties of the kinases over a long period of evolution. The existence of several related eIF2α-specific protein kinases may provide cells with a certain degree of redundancy of function, and this may account for the lack of a strong phenotype in mice that are null for a single member of this kinase family such as PKR (Yang et al. 1995; Abraham et al. 1999). It is uncertain, however, whether the properties of even these four enzymes can account for all the changes in

Table 1. The eIF2α protein kinases

Name	M_r (kDa)	Physiological role	Activators	Inhibitors
HRI	70	Haem-regulated protein synthesis	Iron or haemin deficiencies; heat shock; heavy metals	Adequate haemin levels; normal physiological temperature; hsp90
PKR	62–65	dsRNA-regulated protein synthesis; growth control; apoptosis	dsRNA; structured mRNAs; heparin; PACT/RAX	High concentrations of dsRNA; Alu RNAs; $p58IP^K$; TAR-BP; ribosomes; La antigen; p67
PERK	125	Translational inhibition during the unfolded protein response	Unfolded proteins in the ER	??
GCN2	182	Translational activation of GCN4 synthesis in yeast	Nitrogen starvation; uncharged tRNA	Hsp90

eIF2α phosphorylation that occur under various physiological conditions and it remains possible that other eIF2α kinases still await identification.

2.1 Double-Stranded RNA-Activated Protein Kinase (PKR)

PKR has been the most widely studied mammalian eIF2α kinase in recent years. It has been cloned from human and mouse cells, revealing several features that are characteristic of the protein kinase family in general (Meurs et al. 1990; Icely et al. 1991). In addition to these common features, a "kinase insertion sequence" occurs in PKR between the protein kinase subdomains V and VI, containing the sequence LXIQMXXC. The latter is also found in HRI and GCN2 (Chong et al. 1992) and was originally thought to be involved in substrate recognition. However, more recent studies have established that this sequence is dispensible for interaction with eIF2α and that, at least for PKR, the ability to bind the substrate resides within other parts of the C-terminal half of the protein (Craig et al. 1996; Gale et al. 1996; Sharp et al. 1997, 1998).

PKR binds to double-stranded RNA (dsRNA) with high affinity and is normally dependent on low concentrations of dsRNA for its activation. The structural features required for the interaction between PKR and dsRNA include two dsRNA-binding motifs near the N-terminus of the protein (amino acids 11–77 and 101–167 in human PKR) (reviewed in Clemens 1997; Clemens and Elia 1997). These motifs have features in common with sequences found in several other dsRNA-binding proteins (St Johnston et al. 1992). The solution structure of the dsRNA-binding region of PKR has been analysed by NMR, providing a model for the interaction of this part of the protein with the RNA (Nanduri et al. 1998). However, it is likely that other parts of the PKR molecule, such as a basic domain between amino acids 233 and 271, are also important for its regulation by dsRNA (Rivas et al. 1999).

In addition to synthetic dsRNAs, viral RNAs with extensive secondary structure, such as reovirus S1 mRNA (Henry et al. 1994) and hepatitis δ RNA (Robertson et al. 1996; Circle et al. 1997), are good activators of PKR, presumably because they mimic dsRNA, and it is likely that some highly structured cellular mRNAs also possess this property. Examples of the latter may include transcripts encoding human TNFα (Osman et al. 1999), the dystrophia myotonica protein kinase (Tian et al. 2000) and the translationally controlled growth-regulated protein p23 (U.-A. Bommer and M.J. Clemens, unpubl. data). Evidence that the translation of PKR itself may be impaired by activation of the protein kinase, leading to a self-limiting rate of synthesis of the enzyme (Thomis and Samuel 1992; Barber et al. 1993; Lee et al. 1993), similarly suggests that the mRNA for PKR may be able to activate its own translation product. Different splice variants of the PKR message, containing alternative 5′ untranslated regions (Kawakubo et al. 1999), may show differences in this respect. The kinase can also be activated by the products of transcription of transfected plasmids (Kaufman and Murtha 1987; Kaufman et al. 1989; Choi et al. 1992; Terenzi et al. 1999), and by other molecules such as heparin (Patel et al. 1994), but the molecular basis for this remains unclear. There are also well-characterised RNAs that act as inhibitors of PKR. The best studied are the adenovirus-encoded small RNA VA_I (Mathews and Shenk 1991) and the Epstein-Barr virus-encoded EBER-1 (Sharp et al. 1993; Clemens et al. 1994). There may be equivalent cellular RNAs that fulfill a similar role, with Alu-containing transcripts being good candidates for this function in cells responding to various stresses (Chu et al. 1998).

Activation of PKR usually requires protein dimerisation involving hydrophobic side-chain interactions (Romano et al. 1995; Patel and Sen 1998a), and this probably results in conformational changes which facilitate the phosphorylation of one kinase molecule by its neighbour. This may occur while both monomers are bound to a single stretch of dsRNA. The existence of PKR in multiple forms with varying isoelectric points (Jeffrey et al. 1995) is consistent with the finding that the protein is autophosphorylated at several sites during its activation (Taylor et al. 1996; Romano et al. 1998). Once the enzyme is phosphorylated, its activity against other substrates becomes independent of dsRNA. There is an additional potential mechanism by which PKR may be activated. Two homologous proteins from human and mouse cells, PACT and RAX, have been described which can bind to PKR and activate it in the absence of added dsRNA (Patel and Sen 1998b; Ito et al. 1999b). Again, PKR dimerisation appears to be a prerequisite for this process. Curiously, both PACT and RAX themselves have characteristics of dsRNA-binding proteins, raising the question of the true dsRNA-independence of PKR activation by these factors.

Interestingly, PKR is present not only in the cytoplasm as a ribosome-associated protein but also in the cell nucleus (Jiménez-García et al. 1993; Jeffrey et al. 1995). The majority of this nuclear PKR is in the nucleolus, where it may be associated with nascent ribosomes. The tight association of PKR with ribosomes has an inhibitory effect on the activity of the kinase but both ribosome binding and the inhibition of PKR can be reversed in the presence of dsRNA (Raine et al. 1998; Kumar et al. 1999). This mechanism may restrict the kinase to activation by those cellular RNAs that have only the highest affinities for PKR. There are also a number of protein inhibitors of PKR (Table 1), some of which have been identified by genetic approaches (Tan and Katze 1998), and these proteins may further limit the activity of the enzyme under normal conditions.

The original interest in PKR was due to the involvement of this kinase in the antiviral effects of the interferons (Meurs et al. 1992; Stark et al. 1998). The expression of PKR can be induced several-fold by interferon treatment and the enzyme becomes activated in infected cells, leading to impairment of the overall rate of protein synthesis and a slowing of viral replication (Meurs et al. 1992; Lee and Esteban 1993). It is now clear, however, that PKR has a variety of additional functions which operate in uninfected cells. It has been proposed to play a role in signal transduction, e.g. in response to mitogenic growth factors such as PDGF (Mundschau and Faller 1995) or interleukin-3 (Ito et al. 1994), and there is evidence for the importance of PKR in cell cycle regulation (Zamanian-Daryoush et al. 1999) and in Ca^{2+} storage and signalling (Thomas et al. 1998). The expression of wild-type PKR in yeast inhibits cell proliferation (Chong et al. 1992) and the protein kinase can exert a tumour-suppressing activity (Koromilas et al. 1992b; Meurs et al. 1993). Expression of catalytically inactive mutant forms in 3T3 cells results in a tumourigenic phenotype (Koromilas et al. 1992b; Meurs et al. 1993), probably because of a dominant negative effect of the mutants on the activity of the endogenous wild type PKR. This is consistent with the observation that overexpression of a non-phosphorylatable mutant form of the α subunit of eIF2, in which Ser^{51} has been replaced by Ala, can result in a tumourigenic phenotype (Donzé et al. 1995). Recent studies have established that PKR is also involved in the regulation of apoptosis and this will be discussed in detail in Section 3.2 below.

There is increasing evidence for the phosphorylation of additional substrates by PKR. Such substrates include the NF90 protein, which appears to be a member of the NFAT family of transcription factors (Langland et al. 1999), and DRBP76, an M phase-specific nuclear phosphoprotein (Patel et al. 1999). Another PKR-binding protein, p74, which is closely related in sequence to NF90, has recently been described (Coolidge and Patton 2000). In common with several previously described PKR-associated proteins (Gatignol et al. 1993; Park et al. 1994; Sharp et al. 1998), NF90, DRBP76 and p74 all contain dsRNA-binding motifs. This common feature is likely to constitute the basis for the interactions of these otherwise diverse proteins with PKR. Although not all PKR-interacting proteins are capable of being phosphorylated by the kinase, future experiments will no doubt reveal other true substrates for the enzyme. In particular, the reported ability of PKR to phosphorylate tyrosine as well as serine or threonine residues (Lu et al. 1999) raises the possibility of the enzyme being involved in a much wider range of regulatory activities, which may be especially relevant to its role in signal transduction. Other proteins with transcriptional roles that are regulated directly or indirectly by PKR include the signal transducer and activator of transcription, STAT1 (Wong et al. 1997; Tam et al. 1999), the interferon regulatory factor IRF-1 (Kirchhoff et al. 1995), the tumour suppressor p53 (Yeung and Lau 1998; Cuddihy et al. 1999a, b) and members of the NFκB family (Cheshire et al. 1999; Demarchi et al. 1999). The impaired signalling response to dsRNA treatment observed in PKR-deficient fibroblasts (Kumar et al. 1997) is probably at least partly due to the failure to activate NFκB as a consequence of reduced IκB kinase activity (Zamanian-Daryoush et al. 2000).

2.2 The *Saccharomyces cerevisiae* Protein Kinase GCN2

The *Saccharomyces cerevisiae* enzyme GCN2 was identified as an eIF2α-specific protein kinase following extensive genetic and biochemical investigations into the mechanism underlying the derepression of the transcription factor GCN4 in cells subjected to amino acid starvation (Hinnebusch 1996). Stimulation of GCN2 activity leads to enhanced synthesis of GCN4 at the translational level during nitrogen deficiency (Dever et al. 1992). This is at first sight paradoxical since it involves stimulation of translation of a protein under conditions where overall protein synthesis should be decreased. The mechanism for this requires the presence of upstream open reading frames (ORFs) in the GCN4 mRNA, translation of which attenuates reinitiation at the downstream AUG codon of the GCN4 ORF (Abastado et al. 1991; Pain 1994). Under conditions of amino acid limitation, eIF2α becomes phosphorylated by the activated GCN2 kinase, resulting in a lower level of [eIF2.Met-tRNA$_f$.GTP] ternary complexes, reduced frequency of reinitiation at upstream ORFs 3 and 4 and an increased probability of ribosomes scanning further along the mRNA to reach the downstream GCN4 ORF before reinitiation occurs.

In accordance with the stimulation of GCN2 activity by amino acid starvation, the proximal activator of this protein kinase is thought to be uncharged tRNA. Sequences have been identified in GCN2 which, by virtue of their similarities to histidyl-tRNA synthetases, may be involved in binding uncharged tRNA and these have been shown to be required for the amino acid starvation response (Wek et al. 1989; Ramirez et al. 1992). In addition to these regions, yeast GCN2 has some sequences similar to those of mammalian PKR (Chong et al. 1992) but these probably represent sites required for recognition of the eIF2α substrate or for binding to ribosomes (Ramirez et al. 1991; Marton et al. 1997), rather than being involved with the mechanism of activation of the enzyme per se.

It has been known for many years that mammalian cells also respond to amino acid starvation with increased phosphorylation of eIF2α and downregulation of protein synthesis (Pain 1994). However the enzyme responsible was not identified in these earlier studies. With the recent cloning and characterisation of a mouse homologue of *S. cerevisiae* GCN2 (Berlanga et al. 1999) a likely candidate for the protein kinase that mediates the response to amino acid starvation has been revealed. Nevertheless, it has not yet been formally shown that mammalian GCN2 is activated to phosphorylate eIF2α under these conditions. Although there are several mammalian genes which have upstream ORFs in their mRNAs (Brown et al. 1999; Child et al. 1999; Jagus et al. 1999), none has yet been shown to be upregulated at the translational level in an analogous way to GCN4 in response to nutrient deprivation or other conditions which decrease the activity of eIF2.

2.3
PKR-Like Endoplasmic Reticulum Protein Kinase (PERK)

PERK (also known as PEK) is a recently identified member of the mammalian eIF2α kinase family, which provides yet another link between changes in the intracellular or extracellular environment and the regulation of protein synthesis. It is a type I transmembrane protein, located in the endoplasmic reticulum, which possesses a lumenal domain able to recognise incorrectly folded proteins (Harding et al. 1999). This region of PERK is distinct from the sequences of other eIF2α kinases but bears some similarities to the unfolded protein response protein IRE1B. It transmits signals to a cytoplasmically located protein kinase domain, which does resemble those of PKR, and the other eIF2α kinases. It has been shown that the stress of the presence of unfolded proteins in the endoplasmic reticulum (ER) leads to eIF2α phosphorylation and the downregulation of protein synthesis at the level of initiation (Harding et al. 1999). This regulatory circuit thus provides a feedback mechanism to minimise the accumulation of incorrectly folded, newly synthesised proteins by reducing their rate of synthesis.

PERK is particularly abundant in cells which have a major protein secretory activity, such as the somatostatin-synthesising delta cells of pancreatic islets (Shi et al. 1999). Homologues are also found in other, non-mammalian species. However, the activity of PERK in cell types with relatively little secretory function is minimal (Shi et al. 1998, 1999; also S.J. Morley and M.J. Clemens, unpubl. data). This suggests that its role may be relatively specialised and that it is less likely to be involved in general regulatory events such as cellular growth control. However, this remains to be examined in detail. Unlike PKR and GCN2, PERK/PEK may not associate with ribosomes (Shi et al. 1999).

2.4
Haemin-Regulated Inhibitor (HRI)

HRI (also known as HCR) occurs predominantly, but probably not exclusively, in erythroid cells (Crosby et al. 1994; Mellor et al. 1994). The literature is confusing on the question of the distribution of this enzyme in different tissues since, using Northern and Western blotting analyses, Crosby et al. (1994) were unable to detect expression in non-erythroid cell types. However, rat HRI has been cloned from a brain cDNA library and the corresponding mRNA has been identified in both reticulocytes and psoas muscle (Mellor et al. 1994). Furthermore, whereas globin mRNA was not detectable in lung, heart, liver and kidney, low levels of HRI mRNA were present in these tissues. These results have been confirmed by another study in which mouse HRI protein was purified from liver and 3T3 cells, the corresponding cDNA cloned from a liver library and the encoded mRNA detected in reasonable abundance in liver, kidney and testis (Berlanga et al. 1998).

Although the presence of HRI in erythroid cells clearly provides a means for the co-ordination of haem and globin synthesis during erythropoiesis (Chen et al. 1994), it remains unclear what physiological significance is to be attributed to the expression of HRI, or how the enzyme may be regulated, in non-erythroid cell types. It is possible that this kinase may play a wider role in the regulation of protein synthesis, and could substitute for PKR in cells where activity of the latter has been eliminated by gene deletion or expression of PKR dominant negative mutants. HRI, like PKR, appears to have tyrosine kinase, as well as serine and threonine kinase, activity (Lu et al. 1999).

Activation of HRI is prevented by haemin in vivo and in vitro. Conversely, the enzyme becomes functional when reticulocytes are deficient in iron or haem. In addition, many other stimuli can activate the kinase (Clemens 1996), perhaps providing a basis for its regulation in non-erythroid cells. Again HRI has sequence similarities to other eIF2α kinases, with the closest relationships being to GCN2 and PKR (Chen et al. 1991). One notable feature is a large insert separating the protein kinase domains of HRI into two groups. The mechanism of activation has been the subject of controversy. The data suggest that haemin, and some other porphyrins, promote the formation of intermolecular and/or intramolecular disulphide bonds in HRI, resulting in either covalent homodimerisation (Chen et al. 1989), or heterodimerisation between HRI and the heat shock protein hsp90 (Matts et al. 1992; Méndez et al. 1992; Méndez and De Haro 1994). It is possible that hsp90 is associated with the inactive but not the active form of HRI. Activation is accompanied by autophosphorylation of HRI (Berlanga et al. 1998), as well as phosphorylation of the associated hsp90 (Méndez and De Haro 1994). Involvement of casein kinase-II has also been implicated in HRI activation but the relative contributions of this enzyme and of the autokinase activity of HRI itself have not been clarified.

In both erythroid and non-erythroid cells, activation of HRI by heat shock or by conditions that mimic this stress, such as exposure to heavy metals, provides an alternative pathway to phosphorylate eIF2α and inhibit protein synthesis (Matts and Hurst 1992; Matts et al. 1993). Such regulation is consistent with the interaction of HRI with heat shock proteins via disulphide bond formation and with the ability of heavy metals to affect such bonds (Matts et al. 1991). As well as hsp90, HRI can bind to other proteins involved in cellular stress responses, including members of the hsp70 family, and these could regulate its activation during heat shock. Addition of the heat shock protein hsp70(R) to haemin-deficient reticulocyte lysates reverses the HRI-mediated inhibition of protein synthesis (Gross et al. 1994). There is an interesting parallel here with the regulation of the PERK kinase, described in Section 2.3 above, since proteins that become unfolded and denatured in response to heat or other stresses may also activate HRI and downregulate translation. However, the proposed mechanism is different; the denatured proteins probably bind to members of the hsp70 family in competition with HRI, thus leading to dissociation and activation of the latter (Matts and Hurst 1992; Matts et al. 1993), rather than directly interacting with a domain of the protein kinase.

3 Physiological Regulation of eIF2α Phosphorylation

The protein kinases that phosphorylate eIF2α at amino acid Ser^{51} are strongly conserved throughout the eukaryotic kingdom, suggesting that a large variety of regulatory pathways that affect protein synthesis are likely to involve changes in eIF2α phosphorylation (Clemens 1996). Many physiological stresses which downregulate protein synthesis lead to increases in eIF2α phosphorylation. Examples include heat shock (Scorsone et al. 1987; Duncan and Hershey 1989), virus infection (Chinchar and Dholakia 1989; DeStefano et al. 1990; Huang and Schneider 1990), amino acid or glucose starvation (Scorsone et al. 1987), chronic exposure to alcohol (Lang et al. 1999), ischaemia (Hu and Wieloch 1993; Burda et al. 1994) or recovery after a period of ischaemia (DeGracia et al. 1996; DeGracia et al. 1999; Sullivan et al. 1999). Some of these situations are dealt with in detail in Section 3.1 below. The kinases involved have not always been identified, and although the creation of knock-out mice has made it relatively easy to rule out a role for a particular enzyme, proving involvement of another kinase is more tricky. Using such an approach, it has been demonstrated that PKR is not required for the phosphorylation of eIF2α in brain recovering from ischaemia (DeGracia et al. 1999). The phosphorylation of eIF2α in erythroid cells which are deficient in iron or haem (Farrell et al. 1977; Leroux and London 1982) can also be regarded as a (cell-specific) stress response; in this case, HRI is clearly the kinase responsible.

The phosphorylation of eIF2α is not necessarily restricted to stressful or pathological circumstances since it is becoming apparent that it may well also be involved in the normal regulation of cell growth (Mundschau and Faller

1991; Chong et al. 1992). Changes in the phosphorylation state of eIF2α, or in the activities of the relevant protein kinases, can occur when cell proliferation is stimulated by mitogens and growth factors (Montine and Henshaw 1989; Ito et al. 1994), when mitogenic signalling is inhibited by flavonoids such as genistein (Ito et al. 1999a), or when cell differentiation is induced in several systems (Petryshyn et al. 1984, 1988; Kronfeld-Kinar et al. 1999). Moreover, the association of eIF2α phosphorylation, and involvement of the PKR kinase, in particular, with early events in cell death (apoptosis) (reviewed in Clemens et al. 2000) indicates an even wider role in the control of cellular homeostasis.

Alterations in eIF2α phosphorylation and eIF2 activity in mammalian cells can also result from changes in intracellular and/or extracellular calcium levels (Brostrom and Brostrom 1990; Palfrey and Nairn 1995; Aktas et al. 1998). The mobilisation of sequestered calcium from intracellular stores inhibits translation by a mechanism involving changes in the state of phosphorylation of eIF2α. This phenomenon may provide a basis for the effects of hormones such as vasopressin, epinephrine or angiotensin II on protein synthesis in the liver, via the production of inositol trisphosphate (Kimball and Jefferson 1992; Prostko et al. 1992, 1993). The protein kinase responsible for the Ca^{2+}-dependent regulation of eIF2 activity is most likely PKR (Prostko et al. 1995; Srivastava et al. 1995; Aktas et al. 1998). The latter may be directly activated, or at least sensitised to lower concentrations of dsRNA, by the protein activators PACT/RAX in response to Ca^{2+} mobilisation (and other stresses) (Patel and Sen 1998b; Ito et al. 1999b). However, the more recent characterisation of PERK/PEK as an enzyme regulated by changes in endoplasmic reticulum function (Shi et al. 1998, 1999; Harding et al. 1999) raises the question of whether this kinase may also play a role in Ca^{2+}-mediated effects.

Although it is possible that in some cases alterations in the phosphorylation state of eIF2α and the activity of the eIF2/eIF2B system are consequences of physiological responses, rather than initiating causes, such changes have been demonstrated as necessary for the inhibitory effects on protein synthesis of various cell treatments. The best evidence comes from experiments where the phosphorylation site in eIF2α (Ser^{51}) has been changed to alanine by site-directed mutagenesis (the S51 A mutation). This can result in substantial protection of protein synthesis from inhibition by heat shock, virus infection or plasmid transfection (Davies et al. 1989; Kaufman et al. 1989; Choi et al. 1992; Murtha-Riel et al. 1993; Ramaiah et al. 1994). An endogenous glycoprotein, p67, can also serve to protect eIF2α from phosphorylation (Ray et al. 1992, 1993) and may function to regulate protein synthesis (Wu et al. 1996), for example under conditions such as heat shock or mitosis (Chatterjee et al. 1998; Datta et al. 1999).

3.1
Stress Responses

The regulation of protein synthesis during cellular responses to physiological stress has been reviewed by several authors in recent years (Brostrom and Brostrom 1998; Kaufman 1999; Sheikh and Fornace 1999; Williams 1999). While it is clear that many intracellular signalling pathways are involved (Kleijn et al. 1998; Rhoads 1999), the activation of eIF2α kinases and/or increased phosphorylation of eIF2α are frequently observed phenomena. The relative contributions of these responses to the overall downregulation of protein synthesis, and the mechanisms by which they occur, are not always evident in various systems, and probably vary with the severity of the stress. One intriguing finding is that the phosphorylation of eIF2α can influence the formation of "stress granules" containing untranslated mRNAs. Cells containing S51 A mutant eIF2α fail to assemble these particles, whereas a phosphomimetic mutation (S51D) of the factor induces stress granule formation (Kedersha et al. 1999). However, the mechanism linking the phosphorylation of eIF2α to the sequestration of mRNAs in the granules remains to be elucidated.

3.1.1
Heat Shock

Heat shock is a well-studied example of a condition which causes increased phosphorylation of eIF2α and inhibits eIF2 activity in many cell types (Scorsone et al. 1987; Duncan and Hershey 1989; Hu et al. 1993). There is now a substantial body of evidence indicating the association of various members of the heat shock protein (hsp) family with eIF2α-specific kinases (Matts et al. 1992; Matts and Hurst 1992; Table 2). Such proteins may act as chaperones or direct inhibitors of these kinases. Dissociation of the hsp from the enzyme in response to stress would then lead to activation of the latter. In the case of heat shock, the accumulation of denatured proteins probably provides the signal for this dissociation, with these proteins competing with the kinase for binding to

Table 2. Heat shock proteins and molecular chaperones associated with the eIF2α kinases

Protein kinase	Associated proteins	Function of hsp/chaperone association
HRI	Hsp90; hsp70(R)	Inhibition by formation of heterodimers (inhibition reversed by haem deficiency or heavy metals)
PKR	p58IP^{K}; hsp40; Hsc70	Inhibition by p58^{IPK} (this inhibition is relieved by hsp40)
GCN2	Hsp90	Inhibition (inhibition relieved by uncharged tRNA)

the hsps (Matts and Hurst 1992). An alternative interpretation, suggested for the case of hsp70, is that the protein facilitates the dephosphorylation of eIF2α during the recovery from heat shock, rather than preventing phosphorylation of the initiation factor (Chang et al. 1994).

The identity of the eIF2α kinase responsible for the heat shock response probably depends on the cell type. In reticulocytes there is good evidence for the involvement of HRI (Matts et al. 1992; Matts and Hurst 1992), but in other cell types PKR is the strongest candidate. The latter enzyme appears to be subject to regulation by a complex network of hsps, including the proteins $p58^{IPK}$, hsp40 and Hsc70 (Melville et al. 1997, 1999; Tang et al. 1999). These regulators can modulate PKR activity not only in response to heat shock but following virus infection (Lee et al. 1994) and probably other stresses also. Hsc70 has been further implicated in the regulation of HRI (Uma et al. 1999).

3.1.2
Nutrient Supply

Variations in essential nutrient supply, in particular the availability of amino acids and glucose, can affect protein synthesis at the level of polypeptide chain initiation. Several mechanisms have now been identified, viz. rapid alterations in the extent of eIF2α phosphorylation (Scorsone et al. 1987; Pain 1994; Kimball et al. 1998c), direct effects on eIF2B activity not involving eIF2α phosphorylation (Kimball et al. 1998c; Wang et al. 1998; Shah et al. 1999), and regulation of the activity of the mTOR/$p70^{S6}$ kinase pathway. The last of these affects the phosphorylation state and activity of the eIF4E-binding proteins and/or the protein kinase which phosphorylates ribosomal protein S6 ($p70^{S6k}$) (Fox et al. 1998; Kimball et al. 1998c, 1999; Wang et al. 1998; Xu et al. 1998; Shah et al. 1999). In *S. cerevisiae*, TOR regulates the stability of initiation factor eIF4G, the rate of overall protein synthesis and cell cycle progression (Barbet et al. 1996; Berset et al. 1998). At least some of the effects of amino acid starvation can be mimicked in mammalian cells by inactivating temperature-sensitive aminoacyl-tRNA synthetases at a non-permissive temperature (Clemens et al. 1987; Pollard et al. 1989). The nature of the signal initiating the effects of amino acids is not known, and in mammalian cells the kinase which is responsible for the phosphorylation of eIF2α under starvation conditions has not been identified. The existence of a mammalian homologue of the well-characterised yeast enzyme GCN2 (Berlanga et al. 1999) makes it probable that this kinase is involved in the response to amino acid deprivation. Whether mGCN2 is activated by the accumulation of uncharged tRNA, as appears to be the case for the yeast kinase (Wek et al. 1989; Ramirez et al. 1992), also remains to be established. Earlier studies have indicated that this may not be the case in mammalian cells (Pollard et al. 1989). As with the other eIF2α kinases, yeast GCN2 may be negatively regulated by its association with a heat shock protein, in this case Hsp90, and it has been suggested that uncharged tRNA relieves this effect by displacing the Hsp90 (Donzé and Picard 1999).

In mammalian cells it remains to be shown whether alterations in nutrient availability exert specific effects on the expression of individual genes at the translational level, as in the case of the regulation of yeast GCN4 (Hinnebusch 1990, 1996; Dever et al. 1992; Altmann and Trachsel 1993). It is possible that, under conditions where eIF2α phosphorylation is increased, a GCN4-type induction process can operate in higher cells since there are several examples of polycistronic mammalian mRNAs in which short upstream ORFs precede the main coding sequence (Brown et al. 1999; Child et al. 1999; Jagus et al. 1999). A number of mammalian proteins are known to be induced following amino acid starvation, by mechanisms which at least partially involve post-transcriptional regulation (Bruhat et al. 1997; Aulak et al. 1999). Changes in mRNA stability rather than translation have been implicated in these cases, but a GCN4-type mode of regulation remains an additional possibility.

3.1.3
Regulation by Calcium

As indicated above, alterations in intracellular and extracellular calcium concentrations or pool sizes can affect protein synthesis through changes in the phosphorylation state of eIF2α. In this case, there are additional longer term adaptive responses to calcium changes, which can also be correlated with the regulation of eIF2α phosphorylation. Calcium mobilisation leads to transcriptional induction of genes coding for proteins such as Grp78/BiP, an ER-associated molecule which co-ordinates translation with protein processing and facilitates the subsequent tolerance of protein synthesis to the original inhibitory stress (Brostrom et al. 1990; Laitusis et al. 1999). Grp78 induction itself is independent of eIF2α phosphorylation (Brostrom et al. 1995) but the subsequent adjustment of the initiation rate is associated with a decrease in the phosphorylation of eIF2α (Prostko et al. 1992). This adaptive effect also results in the resistance of protein synthesis to inhibition by other cellular stresses such as elevated temperature or treatment with sodium arsenite (Brostrom et al. 1996).

3.1.4
The Relative Importance of eIF2α Phosphorylation and Other Mechanisms in Stress Responses

Since many of the stresses that activate the phosphorylation of eIF2α also activate other pathways that have the potential to downregulate protein synthesis, it is important to know the relative extent to which each response contributes to the inhibition. Two pathways which have received a great deal of attention in recent years concern the control of the activity and availability, respectively, of initiation factor eIF4E (Sonenberg and Gingras 1998). The former may be regulated by the phosphorylation of eIF4E itself, whereas the amount of eIF4E available for formation of translational initiation complexes can be controlled by the reversible sequestration of the factor by a small family of binding proteins, the 4E-BPs. This process is also regulated by a phosphorylation pathway, directed at the 4E-BPs themselves; thus hypophosphorylated 4E-BPs bind eIF4E more tightly and reduce its potential availability for association with initiation factor eIF4G to form the eIF4F complex (Gingras et al. 1999). It is often proposed that eIF4E is the rate-limiting factor for overall translation (Sonenberg and Gingras 1998), although this has been challenged by others (Rau et al. 1996). Whether or not this is the case, however, there is no doubt that the pathways which regulate eIF4E are responsive to environmental stimuli and cellular stresses.

Interestingly, many of the same conditions that induce eIF2α phosphorylation also cause inhibition of the phosphorylation of the 4E-BPs. Examples are heat shock (Vries et al. 1997), amino acid starvation (Fox et al. 1998; Hara et al. 1998; Wang et al. 1998; Kimball et al. 1999) or treatment with the stress inducer sodium arsenite (Brostrom et al. 1996; Fraser et al. 1999). Paradoxically, some of these treatments activate the phosphorylation of eIF4E itself (Wong et al. 1997; Fraser et al. 1999). However, studies in a number of systems indicate a lack of correlation between changes in eIF4E or 4E-BP phosphorylation and alterations in the level of eIF4F and/or the *overall* rate of protein synthesis (reviewed by S.J. Morley, this Vol.). In contrast, there is evidence for selective effects of eIF4E availability on the relative translation rates of individual mRNAs (Koromilas et al. 1992a; Shantz and Pegg 1994; Kimball et al. 1999). A reasonable working hypothesis that reconciles these observations is that the effects of cellular stresses on protein synthesis as a whole are probably mediated by changes in the activity of eIF2 (either via its phosphorylation or as a result of direct effects on eIF2B function), whereas regulation of eIF4E and 4E-BP phosphorylation, with consequences for the level of active eIF4F, leads to selective alterations in mRNA translation (Fig. 2).

Fig. 2. Mechanisms for regulation of protein synthesis by stress-induced changes in the activity of the translational machinery. Several signal transduction pathways are activated by cellular stresses. Increased activity of one or more eIF2α-specific protein kinases results in phosphorylation of eIF2α and inhibition of eIF2B function, as described in Fig. 1. In addition, impairment of activity of mTOR and $p70^{S6k}$ leads to decreased phosphorylation of the 4E-BPs and ribosomal protein S6 under these conditions (Thomas and Hall 1997). Activation of the stress-regulated p38 MAP kinase pathway can also cause increased phosphorylation of eIF4E. Whereas inhibition of the eIF2/eIF2B system is believed to affect mainly the overall rate of protein synthesis (but see Fig. 3), the changes in eIF4E/4E-BP phosphorylation and $p70^{S6k}$ activity have gene-specific effects affecting the translation of mRNAs with structured 5′ untranslated regions and 5′ terminal oligopyrimidine sequences respectively (Jefferies et al. 1997; Sonenberg and Gingras 1998). For details see the text

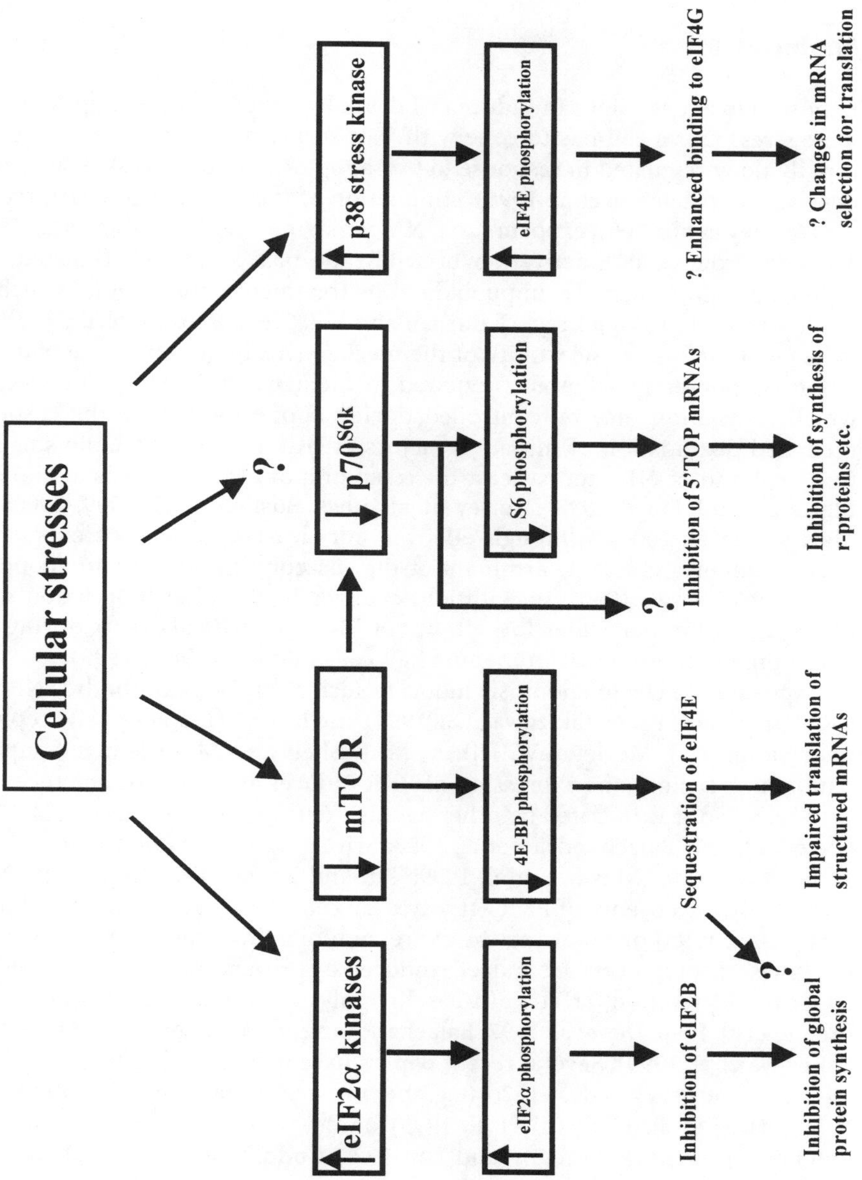

Cellular stresses
eIF2α kinases
mTOR
p70S6k
p38 stress kinase
?
eIF2α phosphorylation
4E-BP phosphorylation
S6 phosphorylation
eIF4E phosphorylation
Inhibition of eIF2B
Sequestration of eIF4E
Inhibition of 5'TOP mRNAs
? Enhanced binding to eIF4G
?
?
Inhibition of global protein synthesis
Impaired translation of structured mRNAs
Inhibition of synthesis of r-proteins etc.
? Changes in mRNA selection for translation

3.2 Apoptosis

Exposure to agents that can induce cell death by apoptosis is perhaps the ultimate stress that a cell has to cope with. Not surprisingly, protein synthesis is rapidly downregulated in response to pro-apoptotic conditions such as serum deprivation (Clemens et al. 1998), stimulation of the Fas (CD95) death receptor (Morley et al. 2000), exposure to DNA damaging agents (Morley et al. 1998; Tee and Proud 2000), activation of wild-type p53 protein (J. Hensold, V.J. Tilleray and M.J. Clemens, unpubl. data) or treatment with cytokines such as TNFα or TRAIL (I.W. Jeffrey, M. Bushell and M.J. Clemens, unpubl. data). What is less expected is the selectivity of the mechanisms by which this downregulation is apparently achieved (reviewed in Clemens et al. 2000). The ways in which translation may be controlled during apoptosis include the caspase-mediated degradation of initiation factors eIF4GI and II, eIF4B, the smallest (p35) subunit of eIF3 and at least the α subunit of eIF2 (Clemens et al. 1998; Marissen and Lloyd 1998; Morley et al. 1998; Bushell et al. 1999, 2000a, b; Marissen et al. 2000). Although eIF4E is not cleaved, its level of expression may be important as a determinant of the susceptibility of cells to apoptosis (Polunovsky et al. 1996). In addition, changes in the phosphorylation state of eIF2α, and in particular the activity of PKR, are critical events during the development of apoptosis in mammalian cells (Tan and Katze 1999).

Exposure of cells to apoptosis inducers such as anti-Fas antibody, TNFα or TRAIL activates PKR (Takizawa et al. 1999) and leads to increased eIF2α phosphorylation (S.J. Morley, I.W. Jeffrey, M. Bushell and M.J. Clemens, unpubl. data). Furthermore, the expression of PKR induces apoptosis or enhances the process when it is initiated by other agents (Yeung et al. 1996; Der et al. 1997; Lee et al. 1997; Balachandran et al. 1998; Srivastava et al. 1998; Gil et al. 1999; Jagus et al. 1999; Takizawa et al. 1999; Tan and Katze 1999; Tang et al. 1999). Conversely, the absence of PKR activity, as in cells derived from knockout mice (Yang et al. 1995) or following the expression of a dominant negative form of the kinase, protects cells from death induced by pro-apoptotic conditions such as serum deprivation, TNFα, anti-Fas, lipopolysaccharide or dsRNA treatment (Yeung et al. 1996; Der et al. 1997; Balachandran et al. 1998; Yeung and Lau 1998; Takizawa et al. 1999). Several recent papers have indicated that PKR may have additional targets besides eIF2α (e.g. the transcriptional regulators NFκB and IRF-1) (Der et al. 1997; Gil et al. 1999) or the tumour suppressor and pro-apoptotic protein p53 (Yeung and Lau 1998; Cuddihy et al. 1999a, b) and so the pro-apoptotic effects of the kinase may not necessarily act via eIF2 phosphorylation. However, a note of caution is necessary in interpreting some studies involving dominant negative PKR mutants because such proteins may show gain of function properties that lead to interference with other dsRNA-regulated enzymes (e.g. 2′5′ oligoadenylate synthetase) (Sharp et al. 1995) or other pathways altogether. Nevertheless, expression of either the inhibitor of eIF2α phosphorylation, p67, or the non-phosphorylatable S51A mutant of eIF2α does partially protect cells from apoptosis (Srivastava et al. 1998; Datta and Datta 1999; Gil et al. 1999). Even more strikingly, the phosphomimetic S51D mutation has been reported to be sufficient to induce apoptosis (Srivastava et al. 1998).

The mechanism for a pro-apoptotic effect of eIF2α phosphorylation remains to be established. A change in subcellular localisation of the phosphorylated initiation factor to the nucleus has been described in neurones undergoing recovery from ischaemia, a condition which is associated with increased apoptosis (DeGracia et al. 1997), but the significance of this is unknown. Several studies suggest that changes in gene expression can occur in response to the activation of PKR during apoptosis. PKR-dependent effects ranging from generalised RNA degradation (Yeung et al. 1996) to induction of Fas mRNA (Der et al. 1997; Balachandran et al. 1998; Gil et al. 1999; Takizawa et al. 1999) and Fas receptor expression (Donzé et al. 1999) have been described, but it is not clear whether these events require the enhanced phosphorylation of eIF2α. Another idea which has been mooted is that increased translation of pro-apoptotic proteins may occur by a GCN4-type mechanism, under conditions where general protein synthesis is inhibited by phosphorylation of eIF2α. Proteins such as Bax and perhaps Fas may be regulated in this way since their mRNAs possess short upstream open reading frames (reviewed in Jagus et al. 1999) and some evidence consistent with this model has been presented (Balachandran et al. 1998). Other candidates for upregulation in response to PKR activation include components of the death-inducing signalling complex (DISC) that are required for cellular responsiveness to dsRNA or anti-Fas stimulation (Balachandran et al. 1998, 2000; Takizawa et al. 1999). Conversely, simply inhibiting the overall rate of protein synthesis, as may occur if eIF2α becomes highly phosphorylated, may have a pro-apoptotic effect since preventing new gene expression at the transcriptional or translational level can result in rapid apoptosis in several cell types (especially transformed cell lines) (Martin and Green 1995; Clemens et al. 1998). This may be caused by the rapid turnover of anti-apoptotic proteins which normally prevent the cell from entering into a "default" state of apoptosis. Cells induced to express dominant negative PKR have higher levels of the anti-apoptotic protein Bcl-2 (and its mRNA) than control cells (Balachandran et al. 1998).

Recently, a further twist in the story of the relationship between eIF2α phosphorylation and apoptosis has been the demonstration that eIF2α is itself a substrate for caspase-mediated cleavage at a site close to the C-terminus (Satoh et al. 1999; Bushell et al. 2000b; Marissen et al. 2000). The significance of this is unclear since only a small fraction of the endogenous protein is cleaved, and the effect is delayed relative to other apoptosis-associated events which affect the protein synthesis machinery. However, the slightly smaller cleavage fragment, even though phosphorylated by PKR, may relieve the inhibition of protein synthesis (Satoh et al. 1999). This could arise because the cleaved protein loses its dependence on eIF2B for guanine nucleotide exchange, causing the latter process to become insensitive to the phosphorylation status of the α subunit (Marissen et al. 2000). If such an effect is operative in vivo it could thus reverse the inhibition of protein synthesis caused by eIF2α phosphorylation, provided that the extent of accumulation of the cleavage fragment is sufficient for this gain of function to acquire physiological significance.

4 What Are the Consequences of eIF2α Phosphorylation?

As summarised in Fig. 3, a variety of regulatory consequences may occur following the phosphorylation of eIF2α at amino acid Ser^{51} (or the equivalent residue in other species). We are familiar with the concept that such phosphorylation leads to the downregulation of overall protein synthesis, and this may

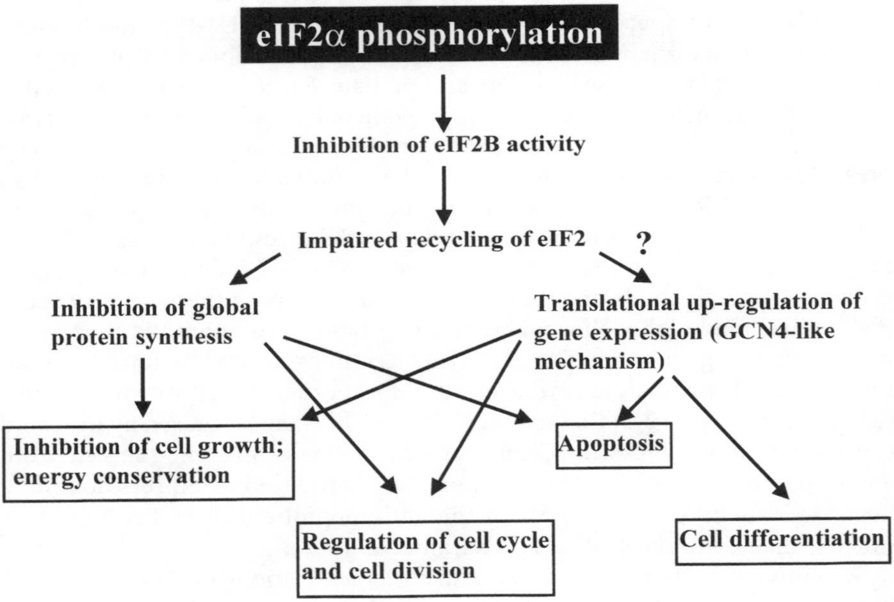

Fig. 3. Possible regulatory consequences of the phosphorylation of eIF2α. Phosphorylation of eIF2α on Ser^{51} (or the equivalent residue in various species) inhibits eIF2B function and the recycling of eIF2 between successive rounds of polypeptide chain initiation (see Fig. 1). This can result in inhibition of global protein synthesis in the cell, leading to impaired cell growth, conservation of energy (due to decreased demand for ATP and GTP for chain elongation), inhibition of cell division (due to failure of cells to acquire the necessary mass) and induction of apoptosis (perhaps as a result of the turnover of anti-apoptotic proteins). In addition, it is possible that the translation of some proteins may be enhanced, by analogy with the mechanism of upregulation of GCN4 synthesis in *S. cerevisiae* (Hinnebusch 1996). This could also contribute to phenotypic changes in the cell, including regulation of passage through the cell cycle, induction of apoptosis or promotion of cell differentiation

indeed be important in cells undergoing responses to stress that require the conservation of energy, the cessation of proliferation or the inhibition of protein synthesis per se (e.g. in erythroid cells deprived of iron, when globin synthesis must not outstrip the availability of haem). The phosphorylation of eIF2α as a means of restricting protein synthesis is probably also important in cells undergoing differentiation or entering apoptosis in response to various external or internal stimuli. In this context, the over-expression of eIF2α in certain tumours (reviewed in Clemens and Bommer 1999) and the correlation of biological aggressiveness of the tumour with the level of expression of the factor (as well as that of eIF4E, Wang et al. 1999) suggest that the regulation of eIF2 may be critical for growth control. This is consistent with the observation that constitutive expression of a non-phosphorylatable eIF2α in 3T3 cells can confer a tumourigenic phenotype on these cells (Donzé et al. 1995).

As well as the global effects on protein synthesis, however, it is possible that there may be paradoxical upregulation of the expression of certain proteins at the translational level following the phosphorylation of eIF2α (Fig. 3). The model for this type of control is GCN4, where the presence of upstream ORFs conveys the ability of the downstream coding region to be translated at an increased rate when eIF2/2B activity is restricted (Hinnebusch 1996). This may occur particularly under conditions where the phosphorylation of eIF2α is only sufficient to cause partial (or no) inhibition of overall protein synthesis. Although no case of this phenomenon has yet emerged in the mammalian field, there are several examples of mRNAs containing upstream ORFs (e.g. Brown et al. 1999; Child et al. 1999; Jagus et al. 1999). Many of these encode proteins of importance in the regulation of cell growth or apoptosis. Thus the effects of the level of eIF2α and of the activities of the kinases which phosphorylate it on cell proliferation or survival could also be mediated by changes in the expression of specific gene products.

An important point to note is that the crucial parameter in mediating the above effects is probably not the absolute level of eIF2α phosphorylation but rather the ratio of the phosphorylated protein to the guanine nucleotide exchange factor eIF2B (Fig. 1). Thus, small changes in eIF2α phosphorylation may have a large impact on total and/or gene-specific protein synthesis in cells with a low eIF2B content, whereas significant levels of eIF2α kinases and of the phosphorylation of their substrate might be tolerated in cells where the eIF2B concentration is higher. Relatively few studies where changes in eIF2α phosphorylation have been reported have also measured the molar ratio of the two factors.

Whatever the functional importance of eIF2α phosphorylation to the cell, the absence of naturally occurring mutations in which Ser^{51} is replaced by other amino acids suggests that selection operates against the loss of a reversibly phosphorylatable residue at this position. Furthermore, the perfect conservation of the amino acid sequence NIEGMILLSEL[S]RRRIRSI around the site of phosphorylation (in square brackets) in eIF2α proteins from very divergent species (Qu and Cavener 1994) supports the idea that this region plays a critical role in eIF2 function. The nature of such function(s) and its significance for cell behaviour will no doubt remain the subject of intensive research for several years to come.

5 Summary

The α subunit of polypeptide chain initiation factor eIF2 can be phosphorylated by a number of related protein kinases which are activated in response to cellular stresses. Physiological conditions which result in eIF2α phosphorylation include virus infection, heat shock, iron deficiency, nutrient deprivation, changes in intracellular calcium, accumulation of unfolded or denatured proteins and the induction of apoptosis. Phosphorylated eIF2 acts as a dominant inhibitor of the guanine nucleotide exchange factor eIF2B and prevents the recycling of eIF2 between successive rounds of protein synthesis. Extensive phosphorylation of eIF2α and strong inhibition of eIF2B activity can result in the downregulation of the overall rate of protein synthesis; less marked changes may lead to alterations in the selective translation of alternative open reading frames in polycistronic mRNAs, as demonstrated in yeast. These mechanisms can provide a signal transduction pathway linking eukaryotic cellular stress responses to alterations in the control of gene expression at the translational level.

Acknowledgements. I wish to thank Drs Jenny Pain, Simon Morley, Jack Hensold, Ian Jeffrey and Martin Bushell and all colleagues in my research group for stimulating discussions. Research in my laboratory is supported by grants from the Wellcome Trust (grants 043488 and 056778), the Leukaemia Research Fund, the Cancer Prevention Research Trust and Glaxo-Wellcome.

References

Abastado J-P, Miller PF, Jackson BM, Hinnebusch AG (1991) Suppression of ribosomal reinitiation at upstream open reading frames in amino acid-starved cells forms the basis for *GCN4* translational control. Mol Cell Biol 11:486–496

Abraham N, Stojdl DF, Duncan PI, Méthot N, Ishii T, Dubé M, Vanderhyden BC, Atkins HL, Gray DA, McBurney MW, Koromilas AE, Brown EG, Sonenberg N, Bell JC (1999) Characterization of transgenic mice with targeted disruption of the catalytic domain of the double-stranded RNA-dependent protein kinase, PKR. J Biol Chem 274:5953–5962

Aktas H, Flückiger R, Acosta JA, Savage JM, Palakurthi SS, Halperin JA (1998) Depletion of intracellular Ca^{2+} stores, phosphorylation of eIF2α, and sustained inhibition of translation initiation mediate the anticancer effects of clotrimazole. Proc Natl Acad Sci USA 95:8280–8285

Altmann M, Trachsel H (1993) Regulation of translation initiation and modulation of cellular physiology. Trends Biochem Sci 18:429–432

Asano K, Krishnamoorthy T, Phan L, Pavitt GD, Hinnebusch AG (1999) Conserved bipartite motifs in yeast eIF5 and eIF2Bî, GTPase-activating and GDP-GTP exchange factors in translation initiation, mediate binding to their common substrate eIF2. EMBO J 18:1673–1688

Aulak KS, Mishra R, Zhou LY, Hyatt SL, De Jonge W, Lamers W, Snider M, Hatzoglou M (1999) Post-transcriptional regulation of the arginine transporter Cat-1 by amino acid availability. J Biol Chem 274:30424–30432

Balachandran S, Kim CN, Yeh WC, Mak TW, Bhalla K, Barber GN (1998) Activation of the dsRNA-dependent protein kinase, PKR, induces apoptosis through FADD-mediated death signaling. EMBO J 17:6888–6902

Balachandran S, Roberts PC, Kipperman T, Bhalla KN, Compans RW, Archer DR, Barber GN (2000) Alpha/beta interferons potentiate virus-induced apoptosis through activation of the FADD/Caspase-8 death signaling pathway. J Virol 74:1513–1523

Barber GN, Wambach M, Wong M-L, Dever TE, Hinnebusch AG, Katze MG (1993) Translational regulation by the interferon-induced double-stranded-RNA-activated 68-kDa protein kinase. Proc Natl Acad Sci USA 90:4621–4625

Barbet NC, Schneider U, Helliwell SB, Stansfield I, Tuite MF, Hall MN (1996) TOR controls translation initiation and early G1 progression in yeast. Mol Biol Cell 7:25–42

Berlanga JJ, Herrero S, De Haro C (1998) Characterization of the hemin-sensitive eukaryotic initiation factor 2α kinase from mouse nonerythroid cells. J Biol Chem 273:32340–32346

Berlanga JJ, Santoyo J, De Haro C (1999) Characterization of a mammalian homolog of the GCN2 eukaryotic initiation factor 2α kinase. Eur J Biochem 265:754–762

Berset C, Trachsel H, Altmann M (1998) The TOR (target of rapamycin) signal transduction pathway regulates the stability of translation initiation factor eIF4G in the yeast *Saccharomyces cerevisiae*. Proc Natl Acad Sci USA 95:4264–4269

Brostrom CO, Brostrom MA (1990) Calcium-dependent regulation of protein synthesis in intact mammalian cells. Annu Rev Physiol 52:577–590

Brostrom CO, Brostrom MA (1998) Regulation of translation initiation during cellular responses to stress. Prog Nucleic Acid Res Mol Biol 58:79–125

Brostrom MA, Cade C, Prostko CR, Gmitter-Yellen D, Brostrom CO (1990) Accommodation of protein synthesis to chronic deprivation of intracellular sequestered calcium. A putative role for GRP78. J Biol Chem 265:20539–20546

Brostrom MA, Prostko CR, Gmitter D, Brostrom CO (1995) Independent signaling of *grp78* gene transcription and phosphorylation of eukaryotic initiation factor 2α by the stressed endoplasmic reticulum. J Biol Chem 270:4127–4132

Brostrom CO, Prostko CR, Kaufmann RJ, Brostrom MA (1996) Inhibition of translational initiation by activators of the glucose-regulated stress protein and heat shock protein stress response systems – role of the interferon-inducible double-stranded RNA-activated eukaryotic initiation factor 2α kinase. J Biol Chem 271:24995–25002

Brown CY, Mize GJ, Pineda M, George DL, Morris DR (1999) Role of two upstream open reading frames in the translational control of oncogene mdm2. Oncogene 18:5631–5637

Bruhat A, Jousse C, Wang XZ, Ron D, Ferrara M, Fafournoux P (1997) Amino acid limitation induces expression of CHOP, a CCAAT/enhancer binding protein-related gene, at both transcriptional and post-transcriptional levels. J Biol Chem 272:17588–17593

Burda J, Martín ME, García A, Alcázar A, Fando JL, Salinas M (1994) Phosphorylation of the α subunit of initiation factor 2 correlates with the inhibition of translation following transient cerebral ischaemia in the rat. Biochem J 302:335–338

Bushell M, McKendrick L, Jänicke RU, Clemens MJ, Morley SJ (1999) Caspase-3 is necessary and sufficient for cleavage of protein synthesis eukaryotic initiation factor 4G during apoptosis. FEBS Lett 451:332–336

Bushell M, Poncet D, Marissen WE, Flotow H, Lloyd RE, Clemens MJ, Morley SJ (2000a) Cleavage of polypeptide chain initiation factor eIF4GI during apoptosis: Characterisation of an internal fragment generated by caspase-3-mediated cleavage. Cell Death Differ 7:628–636

Bushell M, Wood W, Clemens MJ, Morley SJ (2000b) Changes in integrity and association of eukaryotic protein synthesis initiation factors during apoptosis. Eur J Biochem 267:1083–1091

Chang GC, Liu R, Panniers R, Li GC (1994) Rat fibroblasts transfected with the human 70-kDa heat shock gene exhibit altered translation and eukaryotic initiation factor 2 alpha phosphorylation following heat shock. Int J Hypertherm 10:325–337

Chatterjee M, Chatterjee N, Datta R, Datta B, Gupta NK (1998) Expression and activity of p67 are induced during heat shock. Biochem Biophys Res Commun 249:113–117
Chen JJ, Yang JM, Petryshyn R, Kosower N, London IM (1989) Disulfide bond formation in the regulation of eIF-2α kinase by heme. J Biol Chem 264:9559–9564
Chen JJ, Throop MS, Gehrke L, Kuo I, Pal JK, Brodsky M, London IM (1991) Cloning of the cDNA of the heme-regulated eukaryotic initiation factor 2α (eIF-2α) kinase of rabbit reticulocytes: homology to yeast GCN2 protein kinase and human double-stranded-RNA-dependent eIF-2α kinase. Proc Natl Acad Sci USA 88:7729–7733
Chen JJ, Crosby JS, London IM (1994) Regulation of heme-regulated eIF-2α kinase and its expression in erythroid cells. Biochimie 76:761–769
Cheshire JL, Williams BRG, Baldwin AS Jr (1999) Involvement of double-stranded RNA-activated protein kinase in the synergistic activation of nuclear factor-kappaB by tumor necrosis factor-α and gamma-interferon in preneuronal cells. J Biol Chem 274:4801–4806
Child SJ, Miller MK, Geballe AP (1999) Translational control by an upstream open reading frame in the HER-2/neu transcript. J Biol Chem 274:24335–24341
Chinchar VG, Dholakia JN (1989) Frog virus 3-induced translational shut-off: Activation of an eIF-2 kinase in virus-infected cells. Virus Res 14:207–224
Choi S-Y, Scherer BJ, Schnier J, Davies MV, Kaufman RJ, Hershey JWB (1992) Stimulation of protein synthesis in COS cells transfected with variants of the α-subunit of initiation factor eIF-2. J Biol Chem 267:286–293
Chong KL, Feng L, Schappert K, Meurs E, Donahue TF, Friesen JD, Hovanessian AG, Williams BRG (1992) Human p68 kinase exhibits growth suppression in yeast and homology to the translational regulator *GCN2*. EMBO J 11:1553–1562
Chu WM, Ballard R, Carpick BW, Williams BRG, Schmid CW (1998) Potential Alu function: Regulation of the activity of double-stranded RNA-activated kinase PKR. Mol Cell Biol 18:58–68
Circle DA, Neel OD, Robertson HD, Clarke PA, Mathews MB (1997) Surprising specificity of PKR binding to delta agent genomic RNA. RNA Publ RNA Soc 3:438–448
Clemens MJ (1996) Protein kinases that phosphorylate eIF2 and eIF2B, and their role in eukaryotic cell translational control. In: Hershey JWB, Mathews MB, Sonenberg N (eds) Translational control. Cold Spring Harbor Laboratory Press, Cold Spring Harbor, pp 139–172
Clemens MJ (1997) PKR – a protein kinase regulated by double-stranded RNA. Int J Biochem Cell Biol 29:945–949
Clemens MJ, Bommer UA (1999) Translational control: the cancer connection. Int J Biochem Cell Biol 31:1–23
Clemens MJ, Elia A (1997) The double-stranded RNA-dependent protein kinase PKR: structure and function. J Interferon Cytokine Res 17:503–524
Clemens MJ, Galpine A, Austin SA, Panniers R, Henshaw EC, Duncan R, Hershey JWB, Pollard JW (1987) Regulation of polypeptide chain initiation in Chinese hamster ovary cells with a temperature-sensitive leucyl-tRNA synthetase. J Biol Chem 262:767–771
Clemens MJ, Laing K, Jeffrey IW, Schofield A, Sharp TV, Elia A, Matys V, James MC, Tilleray VJ (1994) Regulation of the interferon-inducible eIF-2α protein kinase by small RNAs. Biochimie 76:770–778
Clemens MJ, Bushell M, Morley SJ (1998) Degradation of eukaryotic polypeptide chain initiation factor (eIF) 4G in response to induction of apoptosis in human lymphoma cell lines. Oncogene 17:2921–2931
Clemens MJ, Bushell M, Jeffrey IW, Pain VM, Morley SJ (2000) Translation initiation factor modifications and the regulation of protein synthesis in apoptotic cells. Cell Death Differ 7:603–615
Coolidge CJ, Patton JG (2000) A new double-stranded RNA-binding protein that interacts with PKR. Nucleic Acids Res 28:1407–1417
Craig AWB, Cosentino GP, Donzé O, Sonenberg N (1996) The kinase insert domain of interferon-induced protein kinase PKR is required for activity but not for interaction with the pseudosubstrate K3L. J Biol Chem 271:24526–24533
Crosby JS, Lee K, London IM, Chen J-J (1994) Erythroid expression of the heme-regulated eIF-2α kinase. Mol Cell Biol 14:3906–3914

Cuddihy AR, Li SY, Tam NWN, Wong AHT, Taya Y, Abraham N, Bell JC, Koromilas AE (1999a) Double-stranded-RNA-activated protein kinase PKR enhances transcriptional activation by tumor suppressor p53. Mol Cell Biol 19:2475–2484

Cuddihy AR, Wong AHT, Tam NWN, Li SY, Koromilas AE (1999b) The double-stranded RNA activated protein kinase PKR physically associates with the tumor suppressor p53 protein and phosphorylates human p53 on serine 392 in vitro. Oncogene 18:2690–2702

Datta B, Datta R (1999) Induction of apoptosis due to lowering the level of eukaryotic initiation factor 2-associated protein, p67, from mammalian cells by antisense approach. Exp Cell Res 246:376–383

Datta B, Datta R, Mukherjee S, Zhang ZL (1999) Increased phosphorylation of eukaryotic initiation factor 2α at the G_2/M boundary in human osteosarcoma cells correlates with deglycosylation of p67 and a decreased rate of protein synthesis. Exp Cell Res 250:223–230

Davies MV, Furtado M, Hershey JWB, Thimmappaya B, Kaufman RJ (1989) Complementation of adenovirus virus-associated RNA I gene deletion by expression of a mutant eukaryotic translation initiation factor. Proc Natl Acad Sci USA 86:9163–9167

De Haro C, Méndez R, Santoyo J (1996) The eIF-2α kinases and the control of protein synthesis. FASEB J 10:1378–1387

DeGracia DJ, Neumar RW, White BC, Krause GS (1996) Global brain ischemia and reperfusion: Modifications in eukaryotic initiation factors associated with inhibition of translation initiation. J Neurochem 67:2005–2012

DeGracia DJ, Sullivan JM, Neumar RW, Alousi SS, Hikade KR, Pittman JE, White BC, Rafols JA, Krause GS (1997) Effect of brain ischemia and reperfusion on the localization of phosphorylated eukaryotic initiation factor 2α. J Cereb Blood Flow Metab 17:1291–1302

DeGracia DJ, Adamczyk S, Folbe AJ, Konkoly LL, Pittman JE, Neumar RW, Sullivan JM, Scheuner D, Kaufman RJ, White BC, Krause GS (1999) Eukaryotic initiation factor 2α kinase and phosphatase activity during postischemic brain reperfusion. Exp Neurol 155:221–227

Demarchi F, Gutierrez MI, Giacca M (1999) Human immunodeficiency virus type 1 Tat protein activates transcription factor NF-kappaB through the cellular interferon-inducible, double-stranded RNA-dependent protein kinase, PKR. J Virol 73:7080–7086

Der SD, Yang YL, Weissmann C, Williams BRG (1997) A double-stranded RNA-activated protein kinase-dependent pathway mediating stress-induced apoptosis. Proc Natl Acad Sci USA 94:3279–3283

DeStefano J, Olmsted E, Panniers R, Lucas-Lenard J (1990) The α subunit of eucaryotic initiation factor 2 is phosphorylated in mengovirus-infected mouse L cells. J Virol 64:4445–4453

Dever TE, Feng L, Wek RC, Cigan AM, Donahue TF, Hinnebusch AG (1992) Phosphorylation of initiation factor 2α by protein kinase GCN2 mediates gene-specific translational control of *GCN4* in yeast. Cell 68:585–596

Donzé O, Picard D (1999) Hsp90 binds and regulates the ligand-inducible α subunit of eukaryotic translation initiation factor kinase Gcn2. Mol Cell Biol 19:8422–8432

Donzé O, Jagus R, Koromilas AE, Hershey JWB, Sonenberg N (1995) Abrogation of translation initiation factor eIF-2 phosphorylation causes malignant transformation of NIH 3T3 cells. EMBO J 14:3828–3834

Donzé O, Dostie J, Sonenberg N (1999) Regulatable expression of the interferon-induced double-stranded RNA dependent protein kinase PKR induces apoptosis and Fas receptor expression. Virology 256:322–329

Duncan RF, Hershey JWB (1989) Protein synthesis and protein phosphorylation during heat stress, recovery and adaptation. J Cell Biol 109:1467–1481

Farrell PJ, Balkow K, Hunt T, Jackson RJ, Trachsel H (1977) Phosphorylation of initiation factor eIF-2 and control of reticulocyte protein synthesis. Cell 11:187–200

Fox HL, Pham PT, Kimball SR, Jefferson LS, Lynch CJ (1998) Amino acid effects on translational repressor 4 E-BP1 are mediated primarily by L-leucine in isolated adipocytes. Am J Physiol Cell Physiol 275:C1232–C1238

Fraser CS, Pain VM, Morley SJ (1999) Cellular stress in *Xenopus* kidney cells enhances the phosphorylation of eukaryotic translation initiation factor (elF)4E and the association of elF4F with poly(A)-binding protein. Biochem J 342:519–526

Gale M Jr, Tan SL, Wambach M, Katze MG (1996) Interaction of the interferon-induced PKR protein kinase with inhibitory proteins $p58^{IPK}$ and vaccinia virus K3L is mediated by unique domains: implications for kinase regulation. Mol Cell Biol 16:4172–4181

Gatignol A, Buckler C, Jeang K-T (1993) Relatedness of an RNA-binding motif in human immunodeficiency virus type 1 TAR RNA-binding protein TRBP to human P1/dsI kinase and *Drosophila Staufen*. Mol Cell Biol 13:2193–2202

Gil J, Alcamí J, Esteban M (1999) Induction of apoptosis by double-stranded-RNA-dependent protein kinase (PKR) involves the alpha subunit of eukaryotic translation initiation factor 2 and NF-kappaB. Mol Cell Biol 19:4653–4663

Gingras AC, Gygi SP, Raught B, Polakiewicz RD, Abraham RT, Hoekstra MF, Aebersold R, Sonenberg N (1999) Regulation of 4E-BP1 phosphorylation: a novel two-step mechanism. Genes Dev 13:1422–1437

Gross M, Olin A, Hessefort S, Bender S (1994) Control of protein synthesis by hemin. Purification of a rabbit reticulocyte hsp 70 and characterization of its regulation of the activation of the hemin-controlled eIF-2(α) kinase. J Biol Chem 269:22738–22748

Hara K, Yonezawa K, Weng QP, Kozlowski MT, Belham C, Avruch J (1998) Amino acid sufficiency and mTOR regulate p70 S6 kinase and eIF-4E BP1 through a common effector mechanism. J Biol Chem 273:14484–14494

Harding HP, Zhang Y, Ron D (1999) Protein translation and folding are coupled by an endoplasmic-reticulum-resident kinase. Nature 397:271–274

Henry GL, McCormack SJ, Thomis DC, Samuel CE (1994) Mechanism of interferon action. Translational control and the RNA-dependent protein kinase (PKR): Antagonists of PKR enhance the translational activity of mRNAs that include a 161 nucleotide region from reovirus S1 mRNA. J Biol Regul Homeost Agents 8:15–24

Hinnebusch AG (1990) Involvement of an initiation factor and protein phosphorylation in translational control of *GCN4* mRNA. Trends Biochem Sci 15:148–152

Hinnebusch AG (1996) Translational control of GCN4: gene-specific regulation by phosphorylation of eIF2. In: Hershey JWB, Mathews MB, Sonenberg N (eds) Translational control. Cold Spring Harbor Laboratory Press, Plainsview, NY, pp 199–244

Hu BR, Wieloch T (1993) Stress-induced inhibition of protein synthesis initiation: Modulation of initiation factor 2 and guanine nucleotide exchange factor activities following transient cerebral ischemia in the rat. J Neurosci 13:1830–1838

Hu B-R, Yang Y-BO, Wieloch T (1993) Heat-shock inhibits protein synthesis and eIF-2 activity in cultured cortical neurons. Neurochem Res 18:1003–1007

Huang J, Schneider RJ (1990) Adenovirus inhibition of cellular protein synthesis is prevented by the drug 2-aminopurine. Proc Natl Acad Sci USA 87:7115–7119

Icely PL, Gros P, Bergeron JJM, Devault A, Afar DEH, Bell JC (1991) TIK, a novel serine/threonine kinase, is recognized by antibodies directed against phosphotyrosine. J Biol Chem 266: 16073–16077

Ito T, Jagus R, May WS (1994) Interleukin 3 stimulates protein synthesis by regulating double-stranded RNA-dependent protein kinase. Proc Natl Acad Sci USA 91:7455–7459

Ito T, Warnken SP, May WS (1999a) Protein synthesis inhibition by flavonoids: Roles of eukaryotic initiation factor 2α kinases. Biochem Biophys Res Commun 265:589–594

Ito T, Yang ML, May WS (1999b) RAX, a cellular activator for double-stranded RNA-dependent protein kinase during stress signaling. J Biol Chem 274:15427–15432

Jackson RJ, Hunt SL, Reynolds JE, Kaminski A (1995) Cap-dependent and cap-independent translation – Operational distinctions and mechanistic interpretations. Curr Topics Microbiol Immunol 203:1–29

Jagus R, Joshi B, Barber GN (1999) PKR, apoptosis and cancer. Int J Biochem Cell Biol 31:123–138

Jefferies HBJ, Fumagalli S, Dennis PB, Reinhard C, Pearson RB, Thomas G (1997) Rapamycin suppresses 5′TOP mRNA translation through inhibition of $p70^{S6k}$. EMBO J 16:3693–3704

Jeffrey IW, Kadereit S, Meurs EF, Metzger T, Bachmann M, Schwemmle M, Hovanessian AG, Clemens MJ (1995) Nuclear localization of the interferon-inducible protein kinase PKR in human cells and transfected mouse cells. Exp Cell Res 218:17–27

Jiménez-García LF, Green SR, Mathews MB, Spector DL (1993) Organization of the double-stranded RNA-activated protein kinase DAI and virus-associated VA RNA_I in adenovirus- 2-infected HeLa cells. J Cell Sci 106:11–22

Kaufman RJ (1999) Stress signaling from the lumen of the endoplasmic reticulum: coordination of gene transcriptional and translational controls. Genes Dev 13:1211–1233

Kaufman RJ, Murtha P (1987) Translational control mediated by eucaryotic initiation factor-2 is restricted to specific mRNAs in transfected cells. Mol Cell Biol 7:1568–1571

Kaufman RJ, Davies MV, Pathak VK, Hershey JWB (1989) The phosphorylation state of eucaryotic initiation factor 2 alters translational efficiency of specific mRNAs. Mol Cell Biol 9:946–958

Kawakubo K, Kuhen KL, Vessey JW, George CX, Samuel CE (1999) Alternative splice variants of the human PKR protein kinase possessing different 5′-untranslated regions: Expression in untreated and interferon-treated cells and translational activity. Virology 264:106–114

Kedersha NL, Gupta M, Li W, Miller I, Anderson P (1999) RNA-binding proteins TIA-1 and TIAR link the phosphorylation of eIF-2α to the assembly of mammalian stress granules. J Cell Biol 147:1431–1441

Kimball SR (1999) Eukaryotic initiation factor eIF2. Int J Biochem Cell Biol 31:25–29

Kimball SR, Jefferson LS (1992) Regulation of protein synthesis by modulation of intracellular calcium in rat liver. Am J Physiol Endocrinol Metab 263:E958–E964

Kimball SR, Karinch AM, Feldhoff RC, Mellor H, Jefferson LS (1994) Purification and characterization of eukaryotic translational initiation factor eIF-2B from liver. Biochim Biophys Acta Gen Subj 1201:473–481

Kimball SR, Fabian JR, Pavitt GD, Hinnebusch AG, Jefferson LS (1998a) Regulation of guanine nucleotide exchange through phosphorylation of eukaryotic initiation factor eIF2α. J Biol Chem 273:12841–12845

Kimball SR, Heinzinger NK, Horetsky RL, Jefferson LS (1998b) Identification of interprotein interactions between the subunits of eukaryotic initiation factors eIF2 and eIF2B. J Biol Chem 273:3039–3044

Kimball SR, Horetsky RL, Jefferson LS (1998c) Implication of eIF2B rather than eIF4E in the regulation of global protein synthesis by amino acids in L6 myoblasts. J Biol Chem 273:30945–30953

Kimball SR, Shantz LM, Horetsky RL, Jefferson LS (1999) Leucine regulates translation of specific mRNAs in L6 myoblasts through mTOR-mediated changes in availability of eIF4E and phosphorylation of ribosomal protein S6. J Biol Chem 274:11647–11652

Kirchhoff S, Koromilas AE, Schaper F, Grashoff M, Sonenberg N, Hauser H (1995) IRF-1 induced cell growth inhibition and interferon induction requires the activity of the protein kinase PKR. Oncogene 11:439–445

Kleijn M, Scheper GC, Voorma HO, Thomas AAM (1998) Regulation of translation initiation factors by signal transduction. Eur J Biochem 253:531–544

Koromilas AE, Lazaris-Karatzas A, Sonenberg N (1992a) mRNAs containing extensive secondary structure in their 5′ non-coding region translate efficiently in cells overexpressing initiation factor eIF-4 E. EMBO J 11:4153–4158

Koromilas AE, Roy S, Barber GN, Katze MG, Sonenberg N (1992b) Malignant transformation by a mutant of the IFN-inducible dsRNA-dependent protein kinase. Science 257:1685–1689

Kozak M (1999) Initiation of translation in prokaryotes and eukaryotes. Gene 234:187–208

Kronfeld-Kinar Y, Vilchik S, Hyman T, Leibkowicz F, Salzberg S (1999) Involvement of PKR in the regulation of myogenesis. Cell Growth Differ 10:201–212

Kumar A, Yang YL, Flati V, Der S, Kadereit S, Deb A, Haque J, Reis L, Weissmann C, Williams BRG (1997) Deficient cytokine signaling in mouse embryo fibroblasts with a targeted deletion in the PKR gene: Role of IRF-1 and NF-kappaB. EMBO J 16:406–416

Kumar KU, Srivastava SP, Kaufman RJ (1999) Double-stranded RNA-activated protein kinase (PKR) is negatively regulated by 60 S ribosomal subunit protein L18. Mol Cell Biol 19:1116–1125

Laitusis AL, Brostrom MA, Brostrom CO (1999) The dynamic role of GRP78/BiP in the coordination of mRNA translation with protein processing. J Biol Chem 274:486–493

Lang CH, Wu DQ, Frost RA, Jefferson LS, Vary TC, Kimball SR (1999) Chronic alcohol feeding impairs hepatic translation initiation by modulating eIF2 and eIF4E. Am J Physiol Endocrinol Metab 277:E805–E814

Langland JO, Kao PN, Jacobs BL (1999) Nuclear factor-90 of activated T-cells: a double-stranded RNA-binding protein and substrate for the double-stranded RNA-dependent protein kinase, PKR. Biochemistry 38:6361–6368

Lee SB, Esteban M (1993) The interferon-induced double-stranded RNA-activated human p68 protein kinase inhibits the replication of vaccinia virus. Virology 193:1037–1041

Lee SB, Melkova Z, Yan W, Williams BRG, Hovanessian AG, Esteban M (1993) The interferon-induced double-stranded RNA-activated human p68 protein kinase potently inhibits protein synthesis in cultured cells. Virology 192:380–385

Lee SB, Rodríguez D, Rodríguez JR, Esteban M (1997) The apoptosis pathway triggered by the interferon-induced protein kinase PKR requires the third basic domain, initiates upstream of Bcl-2, and involves ICE-like proteases. Virology 231:81–88

Lee TG, Tang N, Thompson S, Miller J, Katze MG (1994) The 58,000-Dalton cellular inhibitor of the interferon-induced double-stranded RNA-activated protein kinase (PKR) is a member of the tetratricopeptide repeat family of proteins. Mol Cell Biol 14:2331–2342

Leroux A, London IM (1982) Regulation of protein synthesis by phosphorylation of eukaryotic initiation factor 2α in intact reticulocytes and reticulocyte lysates. Proc Natl Acad Sci USA 79:2147–2151

Lu JF, O'Hara EB, Trieselmann BA, Romano PR, Dever TE (1999) The interferon-induced double-stranded RNA-activated protein kinase PKR will phosphorylate serine, threonine, or tyrosine at residue 51 in eukaryotic initiation factor 2α. J Biol Chem 274:32198–32203

Marissen WE, Lloyd RE (1998) Eukaryotic translation initiation factor 4G is targeted for proteolytic cleavage by caspase 3 during inhibition of translation in apoptotic cells. Mol Cell Biol 18:7565–7574

Marissen WE, Guo Y, Thomas AA, Matts RL, Lloyd RE (2000) Identification of caspase 3-mediated cleavage and functional alteration of eukaryotic initiation factor 2alpha in apoptosis. J Biol Chem 275:9314–9323

Martin SJ, Green DR (1995) Protease activation during apoptosis: death by a thousand cuts? Cell 82:349–352

Marton MJ, De Aldana CRV, Qiu HF, Chakraburtty K, Hinnebusch AG (1997) Evidence that GCN1 and GCN20, translational regulators of GCN4, function on elongating ribosomes in activation of eIF2α kinase GCN2. Mol Cell Biol 17:4474–4489

Mathews MB, Shenk T (1991) Adenovirus virus-associated RNA and translation control. J Virol 65:5657–5662

Matts RL, Hurst R (1992) The relationship between protein synthesis and heat shock proteins levels in rabbit reticulocyte lysates. J Biol Chem 267:18168–18174

Matts RL, Schatz JR, Hurst R, Kagen R (1991) Toxic heavy metal ions activate the heme-regulated eukaryotic initiation factor-2α kinase by inhibiting the capacity of hemin-supplemented reticulocyte lysates to reduce disulfide bonds. J Biol Chem 266:12695–12702

Matts RL, Xu Z, Pal JK, Chen J-J (1992) Interactions of the heme-regulated eIF-2α kinase with heat shock proteins in rabbit reticulocyte lysates. J Biol Chem 267:18160–18167

Matts RL, Hurst R, Xu Z (1993) Denatured proteins inhibit translation in hemin-supplemented rabbit reticulocyte lysate by inducing the activation of the heme-regulated eIF-2α kinase. Biochemistry 32:7323–7328

Mellor H, Flowers KM, Kimball SR, Jefferson LS (1994) Cloning and characterization of cDNA encoding rat hemin-sensitive initiation factor-2α (eIF-2α) kinase. Evidence for multitissue expression. J Biol Chem 269:10201–10204

Melville MW, Hansen WJ, Freeman BC, Welch WJ, Katze MG (1997) The molecular chaperone hsp40 regulates the activity of $p58^{IPK}$ the cellular inhibitor of PKR. Proc Natl Acad Sci USA 94:97–102

Melville MW, Tan SL, Wambach M, Song J, Morimoto RI, Katze MG (1999) The cellular inhibitor of the PKR protein kinase, $p58^{IPK}$, is an influenza virus-activated co-chaperone that modulates heat shock protein 70 activity. J Biol Chem 274:3797–3803

Meurs E, Chong K, Galabru J, Thomas NSB, Kerr IM, Williams BRG, Hovanessian AG (1990) Molecular cloning and characterization of the human double-stranded RNA-activated protein kinase induced by interferon. Cell 62:379–390

Meurs EF, Watanabe Y, Kadereit S, Barber GN, Katze MG, Chong K, Williams BRG, Hovanessian AG (1992) Constitutive expression of human double-stranded RNA-activated p68 kinase in murine cells mediates phosphorylation of eukaryotic initiation factor 2 and partial resistance to encephalomyocarditis virus growth. J Virol 66:5805–5814

Meurs EF, Galabru J, Barber GN, Katze MG, Hovanessian AG (1993) Tumor suppressor function of the interferon-induced double-stranded RNA-activated protein kinase. Proc Natl Acad Sci USA 90:232–236

Méndez R, De Haro C (1994) Casein kinase II is implicated in the regulation of heme-controlled translational inhibitor of reticulocyte lysates. J Biol Chem 269:6170–6176

Méndez R, Moreno A, De Haro C (1992) Regulation of heme-controlled eukaryotic polypeptide chain initiation factor 2 α-subunit kinase of reticulocyte lysates. J Biol Chem 267:11500–11507

Montine KS, Henshaw EC (1989) Serum growth factors cause rapid stimulation of protein synthesis and dephosphorylation of eIF-2 in serum deprived Ehrlich cells. Biochim Biophys Acta 1014:282–288

Morley SJ, McKendrick L, Bushell M (1998) Cleavage of translation initiation factor 4G (eIF4G) during anti-Fas IgM-induced apoptosis does not require signalling through the p38 mitogen-activated protein (MAP) kinase. FEBS Lett 438:41–48

Morley SJ, Jeffrey IW, Bushell M, Pain VM, Clemens MJ (2000) Differential requirements for caspase-8 activity in the mechanism of phosphorylation of eIF2α, cleavage of eIF4GI and signaling events associated with the inhibition of protein synthesis in apoptotic Jurkat T cells. FEBS Lett 477:229–236

Mundschau LJ, Faller DV (1991) BALB/c-3T3 fibroblasts resistant to growth inhibition by beta interferon exhibit aberrant platelet-derived growth factor, epidermal growth factor, and fibroblast growth factor signal transduction. Mol Cell Biol 11:3148–3154

Mundschau LJ, Faller DV (1995) Platelet-derived growth factor signal transduction through the interferon-inducible kinase PKR. Immediate early gene induction. J Biol Chem 270:3100–3106

Murtha-Riel P, Davies MV, Scherer BJ, Choi S-Y, Hershey JWB, Kaufman RJ (1993) Expression of a phosphorylation-resistant eukaryotic initiation factor 2 α-subunit mitigates heat shock inhibition of protein synthesis. J Biol Chem 268:12946–12951

Nanduri S, Carpick BW, Yang YW, Williams BRG, Qin J (1998) Structure of the double-stranded RNA-binding domain of the protein kinase PKR reveals the molecular basis of its dsRNA-mediated activation. EMBO J 17:5458–5465

Oldfield S, Jones BL, Tanton D, Proud CG (1994) Use of monoclonal antibodies to study the structure and function of eukaryotic protein synthesis initiation factor eIF-2B. Eur J Biochem 221:399–410

Osman F, Jarrous N, Ben-Asouli Y, Kaempfer R (1999) A cis-acting element in the 3′-untranslated region of human TNF-α mRNA renders splicing dependent on the activation of protein kinase PKR. Genes Dev 13:3280–3293

Pain VM (1994) Translational control during amino acid starvation. Biochimie 76:718–728

Pain VM (1996) Initiation of protein synthesis in eukaryotic cells. Eur J Biochem 236:747–771

Palfrey HC, Nairn AC (1995) Calcium-dependent regulation of protein synthesis. Adv Second Messenger Phosphoprotein Res 30:191–223

Park H, Davies MV, Langland JO, Chang H-W, Nam YS, Tartaglia J, Paoletti E, Jacobs BL, Kaufman RJ, Venkatesan S (1994) TAR RNA-binding protein is an inhibitor of the interferon-induced protein kinase PKR. Proc Natl Acad Sci USA 91:4713–4717

Patel RC, Sen GC (1998a) Requirement of PKR dimerization mediated by specific hydrophobic residues for its activation by double-stranded RNA and its antigrowth effects in yeast. Mol Cell Biol 18:7009–7019

Patel RC, Sen GC (1998b) PACT, a protein activator of the interferon-induced protein kinase, PKR. EMBO J 17:4379–4390

Patel RC, Stanton P, Sen GC (1994) Role of the amino-terminal residues of the interferon-induced protein kinase in its activation by double-stranded RNA and heparin. J Biol Chem 269:18593–18598

Patel RC, Vestal DJ, Xu Z, Bandyopadhyay S, Guo WD, Erme SM, Williams BRG, Sen GC (1999) DRBP76, a double-stranded RNA-binding nuclear protein, is phosphorylated by the interferon-induced protein kinase, PKR. J Biol Chem 274:20432–20437

Pavitt GD, Yang WM, Hinnebusch AG (1997) Homologous segments in three subunits of the guanine nucleotide exchange factor eIF2B mediate translational regulation by phosphorylation of eIF2. Mol Cell Biol 17:1298–1313

Pavitt GD, Ramaiah KVA, Kimball SR, Hinnebusch AG (1998) eIF2 independently binds two distinct eIF2B subcomplexes that catalyze and regulate guanine-nucleotide exchange. Genes Dev 12:514–526

Petryshyn R, Chen J-J, London IM (1984) Growth-related expression of a double-stranded RNA-dependent protein kinase in 3T3 cells. J Biol Chem 259:14736–14742

Petryshyn R, Chen J-J, London IM (1988) Detection of activated double-stranded RNA-dependent protein kinase in 3T3-F442 A cells. Proc Natl Acad Sci USA 85:1427–1431

Pollard JW, Galpine AR, Clemens MJ (1989) A novel role for aminoacyl-tRNA synthetases in the regulation of polypeptide chain initiation. Eur J Biochem 182:1–9

Polunovsky VA, Rosenwald IB, Tan AT, White J, Chiang L, Sonenberg N, Bitterman PB (1996) Translational control of programmed cell death: Eukaryotic translation initiation factor 4 E blocks apoptosis in growth-factor-restricted fibroblasts with physiologically expressed or deregulated Myc. Mol Cell Biol 16:6573–6581

Preiss T, Hentze MW (1999) From factors to mechanisms: translation and translational control in eukaryotes. Curr Opin Genet Dev 9:515–521

Prostko CR, Brostrom MA, Malara EM, Brostrom CO (1992) Phosphorylation of eukaryotic initiation factor (eIF) 2α and inhibition of eIF-2B in GH_3 pituitary cells by perturbants of early protein processing that induce GRP78. J Biol Chem 267:16751–16754

Prostko CR, Brostrom MA, Brostrom CO (1993) Reversible phosphorylation of eukaryotic initiation factor 2α in response to endoplasmic reticular signaling. Mol Cell Biochem 127–128:255–265

Prostko CR, Dholakia JN, Brostrom MA, Brostrom CO (1995) Activation of the double-stranded RNA-regulated protein kinase by depletion of endoplasmic reticular calcium stores. J Biol Chem 270:6211–6215

Qu S, Cavener DR (1994) Isolation and characterization of the Drosophila melanogaster eIF-2α gene encoding the alpha subunit of translation initiation factor eIF-2. Gene 140:239–242

Raine DA, Jeffrey IW, Clemens MJ (1998) Inhibition of the double-stranded RNA-dependent protein kinase PKR by mammalian ribosomes. FEBS Lett 436:343–348

Ramaiah KVA, Davies MV, Chen J-J, Kaufman RJ (1994) Expression of mutant eukaryotic initiation factor 2 α subunit (eIF-2α) reduces inhibition of guanine nucleotide exchange activity of eIF-2B mediated by eIF-2α phosphorylation. Mol Cell Biol 14:4546–4553

Ramirez M, Wek RC, Hinnebusch AG (1991) Ribosome association of GCN2 protein kinase, a translational activator of the *GCN4* gene of Saccharomyces cerevisiae. Mol Cell Biol 11:3027–3036

Ramirez M, Wek RC, Vazquez de Aldana CR, Jackson BM, Freeman B, Hinnebusch AG (1992) Mutations activating the yeast eIF-2α kinase GCN2: Isolation of alleles altering the domain related to histidyl-tRNA synthetases. Mol Cell Biol 12:5801–5815

Rau M, Ohlmann T, Morley SJ, Pain VM (1996) A reevaluation of the cap-binding protein, eIF4E, as a rate-limiting factor for initiation of translation in reticulocyte lysate. J Biol Chem 271:8983–8990

Ray MK, Datta B, Chakraborty A, Chattopadhyay A, Meza-Keuthen S, Gupta NK (1992) The eukaryotic initiation factor 2-associated 67-kDa polypeptide (p^{67}) plays a critical role in regulation of protein synthesis initiation in animal cells. Proc Natl Acad Sci USA 89:539–543

Ray MK, Chakraborty A, Datta B, Chattopadhyay A, Saha D, Bose A, Kinzy TG, Wu S, Hileman RE, Merrick WC, Gupta NK (1993) Characteristics of the eukaryotic initiation factor 2 associated 67-kDa polypeptide. Biochemistry 32:5151–5159

Rhoads RE (1999) Signal transduction pathways that regulate eukaryotic protein synthesis. J Biol Chem 274:30337–30340

Rivas C, Gil J, Esteban M (1999) Identification of functional domains of the interferon-induced enzyme PKR in cells lacking endogenous PKR. J Interferon Cytokine Res 19:1229–1236

Robertson HD, Manche L, Mathews MB (1996) Paradoxical interactions between human delta hepatitis agent RNA and the cellular protein kinase PKR. J Virol 70:5611–5617

Romano PR, Green SR, Barber GN, Mathews MB, Hinnebusch AG (1995) Structural requirements for double-stranded RNA binding, dimerization, and activation of the human eIF-2α kinase DAI in *Saccharomyces cerevisiae*. Mol Cell Biol 15:365–378

Romano PR, Garcia-Barrio MT, Zhang XL, Wang QZ, Taylor DR, Zhang F, Herring C, Mathews MB, Qin J, Hinnebusch AG (1998) Autophosphorylation in the activation loop is required for full kinase activity in vivo of human and yeast eukaryotic initiation factor 2α kinases PKR and GCN2. Mol Cell Biol 18:2282–2297

Rowlands AG, Montine KS, Henshaw EC, Panniers R (1988a) Physiological stresses inhibit guanine-nucleotide-exchange factor in Ehrlich cells. Eur J Biochem 175:93–99

Rowlands AG, Panniers R, Henshaw EC (1988b) The catalytic mechanism of guanine nucleotide exchange factor action and competitive inhibition by phosphorylated eucaryotic initiation factor 2. J Biol Chem 263:5526–5533

Satoh S, Hijikata M, Handa H, Shimotohno K (1999) Caspase-mediated cleavage of eukaryotic translation initiation factor subunit 2α. Biochem J 342:65–70

Scorsone KA, Panniers R, Rowlands AG, Henshaw EC (1987) Phosphorylation of eukaryotic initiation factor 2 during physiological stresses which affect protein synthesis. J Biol Chem 262:14538–14543

Shah OJ, Antonetti DA, Kimball SR, Jefferson LS (1999) Leucine, glutamine, and tyrosine reciprocally modulate the translation initiation factors eIF4F and eIF2B in perfused rat liver. J Biol Chem 274:36168–36175

Shantz LM, Pegg AE (1994) Overproduction of ornithine decarboxylase caused by relief of translational repression is associated with neoplastic transformation. Cancer Res 54:2313–2316

Sharp TV, Schwemmle M, Jeffrey I, Laing K, Mellor H, Proud CG, Hilse K, Clemens MJ (1993) Comparative analysis of the regulation of the interferon-inducible protein kinase PKR by Epstein-Barr virus RNAs EBER-1 and EBER-2 and adenovirus VA_1 RNA. Nucleic Acids Res 21:4483–4490

Sharp TV, Xiao Q, Justesen J, Gewert DR, Clemens MJ (1995) Regulation of the interferon-inducible protein kinase PKR and (2′-5′) oligo(adenylate) synthetase by a catalytically inactive PKR mutant through competition for double-stranded RNA binding. Eur J Biochem 230: 97–103

Sharp TV, Witzel JE, Jagus R (1997) Homologous regions of the α subunit of eukaryotic translational initiation factor 2 (eIF2α) and the vaccinia virus K3L gene product interact with the same domain within the dsRNA-activated protein kinase (PKR). Eur J Biochem 250:85–91

Sharp TV, Moonan F, Romashko A, Joshi B, Barber GN, Jagus R (1998) The vaccinia virus E3L gene product interacts with both the regulatory and the substrate binding regions of PKR: Implications for PKR autoregulation. Virology 250:302–315

Sheikh MS, Fornace AJ Jr (1999) Regulation of translation initiation following stress. Oncogene 18:6121–6128

Shi YG, Vattem KM, Sood R, An J, Liang JD, Stramm L, Wek RC (1998) Identification and characterization of pancreatic eukaryotic initiation factor 2 α-subunit kinase, PEK, involved in translational control. Mol Cell Biol 18:7499–7509

Shi YG, An J, Liang JD, Hayes SE, Sandusky GE, Stramm LE, Yang NN (1999) Characterization of a mutant pancreatic eIF-2α kinase, PEK, and co-localization with somatostatin in islet delta cells. J Biol Chem 274:5723–5730

Sonenberg N, Gingras AC (1998) The mRNA 5′ cap-binding protein eIF4E and control of cell growth. Curr Opin Cell Biol 10:268–275

Srivastava SP, Davies MV, Kaufman RJ (1995) Calcium depletion from the endoplasmic reticulum activates the double-stranded RNA-dependent protein kinase (PKR) to inhibit protein synthesis. J Biol Chem 270:16619–16624

Srivastava SP, Kumar KU, Kaufman RJ (1998) Phosphorylation of eukaryotic translation initiation factor 2 mediates apoptosis in response to activation of the double-stranded RNA-dependent protein kinase. J Biol Chem 273:2416–2423

St Johnston D, Brown NH, Gall JG, Jantsch M (1992) A conserved double-stranded RNA-binding domain. Proc Natl Acad Sci USA 89:10979–10983

Stark GR, Kerr IM, Williams BRG, Silverman RH, Schreiber RD (1998) How cells respond to interferons. Annu Rev Biochem 67:227–264

Sudhakar A, Krishnamoorthy T, Jain A, Chatterjee U, Hasnain SE, Kaufman RJ, Ramaiah KVA (1999) Serine 48 in initiation factor 2α (eIF2α) is required for high-affinity interaction between eIF2α(P) and eIF2B. Biochemistry 38:15398–15405

Sullivan JM, Alousi SS, Hikade KR, Bahu NJ, Rafols JA, Krause GS, White BC (1999) Insulin induces dephosphorylation of eukaryotic initiation factor 2α and restores protein synthesis in vulnerable hippocampal neurons after transient brain ischemia. J Cerebral Blood Flow Metab 19:1010–1019

Takizawa T, Tatematsu C, Nakanishi Y (1999) Double-stranded RNA-activated protein kinase (PKR) fused to green fluorescent protein induces apoptosis of human embryonic kidney cells: possible role in the Fas signaling pathway. J Biochem (Tokyo) 125:391–398

Tam NWN, Ishii T, Li SY, Wong AHT, Cuddihy AR, Koromilas AE (1999) Upregulation of STAT1 protein in cells lacking or expressing mutants of the double-stranded RNA-dependent protein kinase PKR. Eur J Biochem 262:149–154

Tan SL, Katze MG (1998) Using genetic means to dissect homologous and heterologous protein-protein interactions of PKR, the interferon-induced protein kinase. Methods 15:207–223

Tan SL, Katze MG (1999) The emerging role of the interferon-induced PKR protein kinase as an apoptotic effector: a new face of death. J Interferon Cytokine Res 19:543–554

Tang NM, Korth MJ, Gale M Jr, Wambach M, Der SD, Bandyopadhyay SK, Williams BRG, Katze MG (1999) Inhibition of double-stranded RNA- and tumor necrosis factor alpha-mediated apoptosis by tetratricopeptide repeat protein and cochaperone $p58^{IPK}$. Mol Cell Biol 19:4757–4765

Taylor DR, Lee SB, Romano PR, Marshak DR, Hinnebusch AG, Esteban M, Mathews MB (1996) Autophosphorylation sites participate in the activation of the double-stranded-RNA-activated protein kinase PKR. Mol Cell Biol 16:6295–6302

Tee AR, Proud CG (2000) DNA-damaging agents cause inactivation of translational regulators linked to mTOR signalling. Oncogene 19:3021–3031

Terenzi F, DeVeer MJ, Ying H, Restifo NP, Williams BRG, Silverman RH (1999) The antiviral enzymes PKR and RNase L suppress gene expression from viral and non-viral based vectors. Nucleic Acids Res 27:4369–4375

Thomas D, Kim HY, Morgan R, Hanley MR (1998) Double-stranded-RNA-activated protein kinase (PKR) regulates Ca^{2+} stores in *Xenopus* oocytes. Biochem J 330:599–603

Thomis DC, Samuel CE (1992) Mechanism of interferon action: Autoregulation of RNA-dependent P1/eIF-2α protein kinase (PKR) expression in transfected mammalian cells. Proc Natl Acad Sci USA 89:10837–10841

Thomas G, Hall MN (1997) TOR signalling and control of cell growth. Curr Opin Cell Biol 9:782–787

Tian B, White RJ, Xia T, Welle S, Turner DH, Mathews MB, Thornton CA (2000) Expanded CUG repeat RNAs form hairpins that activate the double-stranded RNA-dependent protein kinase PKR. RNA 6:79–87

Uma S, Thulasiraman V, Matts RL (1999) Dual role for Hsc70 in the biogenesis and regulation of the heme-regulated kinase of the a subunit of eukaryotic translation initiation factor 2. Mol Cell Biol 19:5861–5871

Vazquez de Aldana CR, Dever TE, Hinnebusch AG (1993) Mutations in the α subunit of eukaryotic translation initiation factor 2 (eIF-2α) that overcome the inhibitory effect of eIF-2α phosphorylation on translation initiation. Proc Natl Acad Sci USA 90:7215–7219

Vries RGJ, Flynn A, Patel JC, Wang XM, Denton RM, Proud CG (1997) Heat shock increases the association of binding protein- 1 with initiation factor 4E. J Biol Chem 272:32779–32784

Wang ST, Rosenwald IB, Hutzler MJ, Pihan GA, Savas L, Chen JJ, Woda BA (1999) Expression of the eukaryotic translation initiation factors 4E and 2α in non-Hodgkin's lymphomas. Am J Pathol 155:247–255

Wang XM, Campbell LE, Miller CM, Proud CG (1998) Amino acid availability regulates p70 S6 kinase and multiple translation factors. Biochem J 334:261–267

Webb BLJ, Proud CG (1997) Eukaryotic initiation factor 2B (eIF2B). Int J Biochem Cell Biol 29:1127–1131

Wek RC, Jackson BM, Hinnebusch AG (1989) Juxtaposition of domains homologous to protein kinases and histidyl-tRNA synthetases in GCN2 protein suggests a mechanism for coupling GCN4 expression to amino acid availability. Proc Natl Acad Sci USA 86:4579–4583

Welsh GI, Proud CG (1993) Glycogen synthase kinase-3 is rapidly inactivated in response to insulin and phosphorylates eukaryotic initiation factor eIF-2B. Biochem J 294:625–629

Welsh GI, Miller CM, Loughlin AJ, Price NT, Proud CG (1998) Regulation of eukaryotic initiation factor eIF2B: glycogen synthase kinase-3 phosphorylates a conserved serine which undergoes dephosphorylation in response to insulin. FEBS Lett 421:125–130

Williams BRG (1999) PKR; a sentinel kinase for cellular stress. Oncogene 18:6112–6120

Wong AHT, Tam NWN, Yang YL, Cuddihy AR, Li SY, Kirchhoff S, Hauser H, Decker T, Koromilas AE (1997) Physical association between STAT1 and the interferon-inducible protein kinase PKR and implications for interferon and double-stranded RNA signaling pathways. EMBO J 16:1291–1304

Wu SY, Rehemtulla A, Gupta NK, Kaufman RJ (1996) A eukaryotic translation initiation factor 2-associated 67 kDa glycoprotein partially reverses protein synthesis inhibition by activated double-stranded RNA-dependent protein kinase in intact cells. Biochemistry 35:8275–8280

Xu G, Kwon G, Marshall CA, Lin TA, Lawrence JC Jr, McDaniel ML (1998) Branched-chain amino acids are essential in the regulation of PHAS-I and p70 S6 kinase by pancreatic β-cells – a possible role in protein translation and mitogenic signaling. J Biol Chem 273:28178–28184

Yang YL, Reis LFL, Pavlovic J, Aguzzi A, Schäfer R, Kumar A, Williams BRG, Aguet M, Weissmann C (1995) Deficient signaling in mice devoid of double-stranded RNA-dependent protein kinase. EMBO J 14:6095–6106

Yeung MC, Lau AS (1998) Tumor suppressor p53 as a component of the tumor necrosis factor-induced, protein kinase PKR-mediated apoptotic pathway in human promonocytic U937 cells. J Biol Chem 273:25198–25202

Yeung MC, Liu J, Lau AS (1996) An essential role for the interferon-inducible, double-stranded RNA-activated protein kinase PKR in the tumor necrosis factor-induced apoptosis in U937 cells. Proc Natl Acad Sci USA 93:12451–12455

Zamanian-Daryoush M, Der SD, Williams BRG (1999) Cell cycle regulation of the double stranded RNA activated protein kinase, PKR. Oncogene 18:315–326

Zamanian-Daryoush M, Mogensen TH, DiDonato JA, Williams BRG (2000) NF-kappaB activation by double-stranded-RNA-activated protein kinase (PKR) is mediated through NF-kappaB-inducing kinase and IkappaB kinase. Mol Cell Biol 20:1278–1290

Elongation Factor-2 Phosphorylation and the Regulation of Protein Synthesis by Calcium

Angus C. Nairn[1], Masayuki Matsushita[2], Kent Nastiuk[4], Atsuko Horiuchi[1], Ken-Ichi Mitsui[1], Yoshio Shimizu[1], and H. Clive Palfrey[3]

1 Introduction

Eukaryotic protein synthesis is a highly regulated process with many of the key proteins being controlled by phosphorylation (for reviews, see Rhoads 1993; Redpath and Proud 1994; Jefferies and Thomas 1996; Merrick and Hershey 1996; Sonenberg and Gingras 1998). Several initiation factors, as well as ribosomal proteins and aminoacyl-tRNA synthetases, are phosphoproteins. While in some cases the physiological role of phosphorylation of these factors is well characterized, in others it remains less clear. In the latter category is elongation factor-2 (eEF2), which catalyzes the translocation of peptidyl-tRNA from the A-site to the P-site on the ribosome. eEF2 is phosphorylated by a highly conserved and specific Ca^{2+}/calmodulin-dependent kinase, termed EF2 kinase (also known as CaM kinase III) at Thr56 and Thr58 (Palfrey and Nairn 1995; Nairn and Palfrey 1996). In vitro, Thr56 is more rapidly phosphorylated by EF2 kinase, and this is sufficient to result in inhibition of protein synthesis. Dephosphorylation of eEF2 by the protein phosphatase 2A (PP2A) results in reactivation of protein synthesis. By implication, similar events in vivo should regulate the rate of elongation and therefore the overall efficiency of protein synthesis.

EF2 kinase is ubiquitously expressed in all mammalian cells, and the phosphorylation of eEF2 is rapidly stimulated by a variety of exogenous ligands, including hormones, growth factors and mitogens, which are known to increase intracellular Ca^{2+}. In many cases, the phosphorylation of eEF2 is transient, reflecting the rapid activation of kinase activity by Ca^{2+} mobilization and subsequent dephosphorylation, most likely by PP2A. Although many of the biochemical details relating to EF2 kinase and the phosphorylation of eEF2 are established, the physiological significance of regulation of polypep-

[1] Laboratory of Molecular and Cellular Neuroscience, Rockefeller University, New York, New York 10021, USA
[2] Okayama University Medical School, Okayama, Japan
[3] Department of Neurobiology, Pharmacology and Physiology, University of Chicago, Chicago, Illinois 60637, USA
[4] Department of Pathology, University of California, Irvine, California 92697, USA

Progress in Molecular and Subcellular Biology, Vol. 27
Robert E. Rhoads (Ed.)

tide chain elongation by Ca^{2+}-dependent phosphorylation of eEF2 remains to be established. However, phosphorylation of eEF2 and the resulting inhibition of protein synthesis is likely to play an important role in the function of Ca^{2+} in all mammalian cells. This chapter reviews recent progress in the study of eEF2 phosphorylation and the role of Ca^{2+} in the regulation of protein synthesis.

2 Phosphorylation of eEF2 and Regulation of Polypeptide Elongation in Vitro

Each cycle of elongation is controlled by the sequential actions of two factors, eEF1 and eEF2, that together with the peptidyl transferase of the ribosome lead to the addition of one amino acid to the elongating polypeptide chain (Merrick and Hershey 1996). eEF2 regulates the final step in the process of elongation in which the peptidyl-tRNA is translocated from the A to the P site on the ribosome. eEF2 contains an N-terminal guanine nucleotide-binding domain that is homologous to p21ras. For each amino acid added to the elongating polypeptide chain, eEF2 binds GTP, the complex attaches to the ribosome, GTP is hydrolyzed, and the eEF2/GDP complex is released from the ribosome. The precise molecular details of how eEF2 catalyzes the translocation of the peptidyl-tRNA are still unclear, but molecular details of the interaction of EF-G (the homologue of EF2 in bacteria) with the prokaryotic ribosome are beginning to emerge.

eEF2 was first identified as a phosphoprotein in two convergent studies. Earlier work had discovered a widely distributed 100 kDa protein that was phosphorylated in a Ca^{2+}/CaM-dependent manner in extracts from various rat tissues and cultured cells (Palfrey 1983; Nairn et al. 1985). Purification and amino acid sequencing of the 100-kDa protein indicated that it was eEF2 (Nairn and Palfrey 1987). Further analysis of the EF2 kinase activity identified it as being highly specific for eEF2, and distinct from the other Ca^{2+}/CaM-dependent protein kinases that were known at that time. In a parallel study, a 100 kDa phosphoprotein was identified in liver extracts and shown to be identical to eEF2 (Ryazanov 1987). Amino acid sequencing indicated that eEF2 was phosphorylated by purified EF2 kinase primarily at Thr56, although Thr58 (and, to a much lesser extent, Thr53) are phosphorylated less efficiently (Nairn and Palfrey 1987; Ovchinnikov et al. 1990; Price et al. 1991). It appears that phosphorylation of Thr58 may be dependent on prior phosphorylation of Thr56 (Redpath et al. 1993), although preliminary studies of eEF2, in which Thr56 was mutated to alanine, suggest that phosphorylation of Thr56 may act in a negative, rather than a positive, manner (K. Nastiuk and A. Nairn, unpubl. results). Notably, Thr56 and Thr58 (but not Thr53) are highly conserved in eEF2 from diverse eukaryotic species, suggesting an important conserved physiological role for eEF2 phosphorylation.

In vitro, phosphorylation of eEF2 results in almost complete inhibition of polypeptide synthesis as measured by poly(U)-directed poly(Phe) synthesis (Nairn and Palfrey 1987; Ryazanov et al. 1988). This effect can be reversed following dephosphorylation with PP2A, the likely physiological eEF2 phosphatase (see below). In addition, studies using reticulocyte lysate indicated that Ca^{2+}-dependent phosphorylation of eEF2 was responsible for the elongation block seen in the presence of okadaic acid (a potent PP2A inhibitor) (Redpath and Proud 1989). It appears that phosphorylation of Thr56 alone is sufficient to inactivate eEF2 in vitro. In addition, Thr56 is the major site phosphorylated in response to physiological stimuli in intact cells (see below).

Although phosphorylation of eEF2 has a profound effect on the activity of the elongation factor, the precise mechanism by which elongation is inhibited is still not well understood. eEF2 function can be broken down into several individual steps that can be analyzed independently. These are (in order): binding of GTP to eEF2, binding of eEF2/GTP to the ribosome, ribosome-dependent GTP hydrolysis, the translocation reaction, dissociation of eEF2/GDP from the ribosome, and GDP/GTP exchange. Some studies of eEF2, the GTP analog GMP-PNP and puromycin, suggested that GTP hydrolysis might not be required for translocation (Merrick and Hershey 1996). However, recent studies of EF-G indicate that it undergoes large changes in conformation during residence on the ribosome, and that GTP hydrolysis is likely to either precede or be coupled to translocation in prokaryotes (Rodnina et al. 1997; Frank and Agrawal 2000; Green 2000; Stark et al. 2000). GTP hydrolysis does not appear to be modulated by eEF2 phosphorylation since no effect of phosphorylation on GTP binding or GTPase activity has been observed (Carlberg et al. 1990; A. Nairn and H.C. Palfrey, unpubl. results), although one study using fluorescence methods observed that phosphorylated eEF2 had an apparent reduced affinity for GTP but not GDP (Dumont-Miscopein et al. 1994). The most likely effect of eEF2 phosphorylation is that it reduces the affinity of eEF2/GTP for the ribosome (Carlberg et al. 1990; Dumont-Miscopein et al. 1994). Moreover, unphosphorylated eEF2 bound to ribosomes is not accessible to EF2 kinase (Nilsson and Nygård 1991; Dumont-Miscopein et al. 1994), suggesting that the region surrounding the phosphorylation site may participate in eEF2-ribosome interaction. However, other studies using the poly(U)-directed poly(Phe) synthesis assay have suggested that phospho-eEF2 is able to compete with dephospho-eEF2 (Ryazanov et al. 1988; Redpath et al. 1993; A. Nairn and H.C. Palfrey, unpubl. results). These results imply that phosphorylated EF2 may interact with ribosomes and act in a "dominant negative" fashion, although the dominant negative model is not supported by studies of reticulocyte lysate (Redpath et al. 1993) or by in vitro assays using puromycin (Ryazanov and Davydova 1989).

An important consideration in these functional studies may be the effect that eEF2 phosphorylation has on polyribosome profiles. Studies using reticulocyte lysate indicate that phosphorylation of eEF2 and elongation block are associated with increased accumulation of polysomes (Redpath and Proud

1989). In addition, in intact reticulocytes treated with a Ca^{2+} ionophore, eEF2 is rapidly phosphorylated and protein synthesis is rapidly inhibited (Mitsui et al. 2001), an effect that is correlated with elongation block and an increase in polysomes (Wong et al. 1991). An analysis of subcellular fractions obtained from intact reticulocytes stimulated with Ca^{2+} ionophore and okadaic acid indicated that the majority of phospho-eEF2 was present in the cytosol, consistent with the idea that phosphorylation reduces the affinity of eEF2 for the ribosome. However, a measurable amount of phospho-eEF2 was found in association with ribosomes (K. Mitsui and A. Nairn, unpubl. results). Taken together, it is not clear how simply reducing the affinity of phospho-eEF2 for the ribosome would account for the profound effects observed on polypeptide elongation and the accumulation of polyribosomes. Notably, the apparent K_d of phospho-eEF2 for ribosomes was 1.7 μM (Carlberg et al. 1990). Since eEF2 is an abundant protein, it is possible that a significant proportion of the phosphorylated factor could still associate with ribosomes.

The N-terminal region of eEF2 contains the GTPase domain common to small molecular weight G proteins, other elongation factors, and initiation factors (Kohno et al. 1986). Thr56 and Thr58 are located in the so-called E (or effector) domain, a short sequence of about 25 amino acids that is conserved only in elongation factors (Fig. 1). Thr56 and Thr58 in eEF2 flank a conserved aspartate residue that is coordinated to the Mg^{2+} ion in both the GTP and GDP forms of, at least, EF-Tu (Sprinzl 1994; Polekhina et al. 1996). By analogy to ras and other related G proteins, the E domain is likely to be involved in mediating the effects of GTP hydrolysis on downstream effector molecules. Comparison of the crystal structures of the GTP- and GDP-bound forms of EF-Tu revealed that a considerable conformational change in the E domain occurs

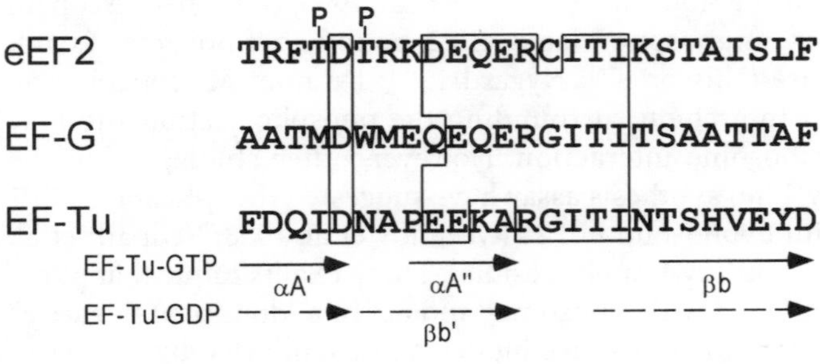

Fig. 1. Comparison of the amino acid sequences of the effector domains of elongation factors. The amino acid sequences of the effector domains of eEF2, EF-G and EF-Tu are illustrated. Residues 53–77 of eEF2 are shown with Thr56 (major) and Thr58 (minor) being the sites phosphorylated by EF2 kinase. Important conserved amino acids in each protein are *boxed*. At the bottom of the illustration, the secondary structures of the relevant regions of EF-Tu/GTP and EF-Tu/GDP are shown. The structure of the effector domain of EF-Tu is altered significantly in these two nucleotide-bound forms. (Polekhina et al. 1996)

upon GTP hydrolysis (Polekhina et al. 1996). A short α-helix in EF-Tu/GTP that follows the conserved aspartate residue undergoes a marked transition to an extended β-sheet conformation in EF-Tu/GDP. Although, experimentally eEF2 phosphorylation does not apparently affect GTP binding or hydrolysis, it seems conceivable that if the E domain of eEF2 resembles that of EF-Tu, phosphorylation of eEF2/GTP could significantly interfere with the ability of the factor to interact constructively with the ribosome. Such an effect might explain both the reduced affinity of phospho-eEF2 for the ribosome, and the ability of bound phospho-eEF2 to act as an inhibitor of polypeptide translocation. Support for the idea that phosphorylation of eEF2 leads to a change in conformation of the E domain comes from the observation that phosphorylation protects Arg66 from proteolysis by trypsin (Nilsson and Nygård 1991). The recent placement of proteins and RNA into a high resolution structure of the bacterial 50 S ribosomal subunit, and the mapping of the binding site for EF-G will hopefully provide additional clues to the potential structural consequence of eEF2 phosphorylation (Wilson and Noller 1998; Ban et al. 1999).

3
Structure and Regulation of EF2 Kinase

EF2 kinase was first purified from reticulocytes and pancreas (Mitsui et al. 1993; Redpath and Proud 1993a). An earlier report that EF2 kinase was identical to the heat shock protein, hsp90, proved to be incorrect (Nygård et al. 1991), although a later report suggested that hsp90 may interact with EF2 kinase under some conditions (Nilsson et al. 1991; Nygård et al. 1991). Based on biochemical studies, together with the cloning and characterization of the recombinant enzyme (see below), EF2 kinase is a monomeric, elongated protein of approximately 95 kDa. Redpath and Proud isolated a cDNA that apparently encoded an EF2 kinase and it was suggested that the cloned EF2 kinase was related in structure to the well-defined family of protein kinases, represented by enzymes such as PKA and PKC (Redpath et al. 1996a). However, based on an analysis of the primary structure of the cloned enzyme and comparison with the crystal structures of PKA and related enzymes, including the Ca^{2+}/CaM-dependent CaM kinase I (Goldberg et al. 1996), it appeared that the protein encoded by the EF2 kinase cDNA had to be completely distinct in structure from the PKA protein kinase family. This was confirmed when Egelhoff and colleagues isolated and characterized a novel myosin heavy chain kinase (MHCK) from dictyostelium containing an approximate 250 amino acid domain that exhibited a high degree of similarity with a central region of the putative EF2 kinase cloned by Proud, but with no resemblance to any domain of PKA (Fig. 2). Additional cloning of EF2 kinase from different species, including rat, mouse, human and *C. elegans* (Ryazanov et al. 1997) and analysis of EST data bases (Ryazanov et al. 1999; M. Matsushita and A. Nairn, unpubl. results), reveal a homologous conserved domain in several additional genes

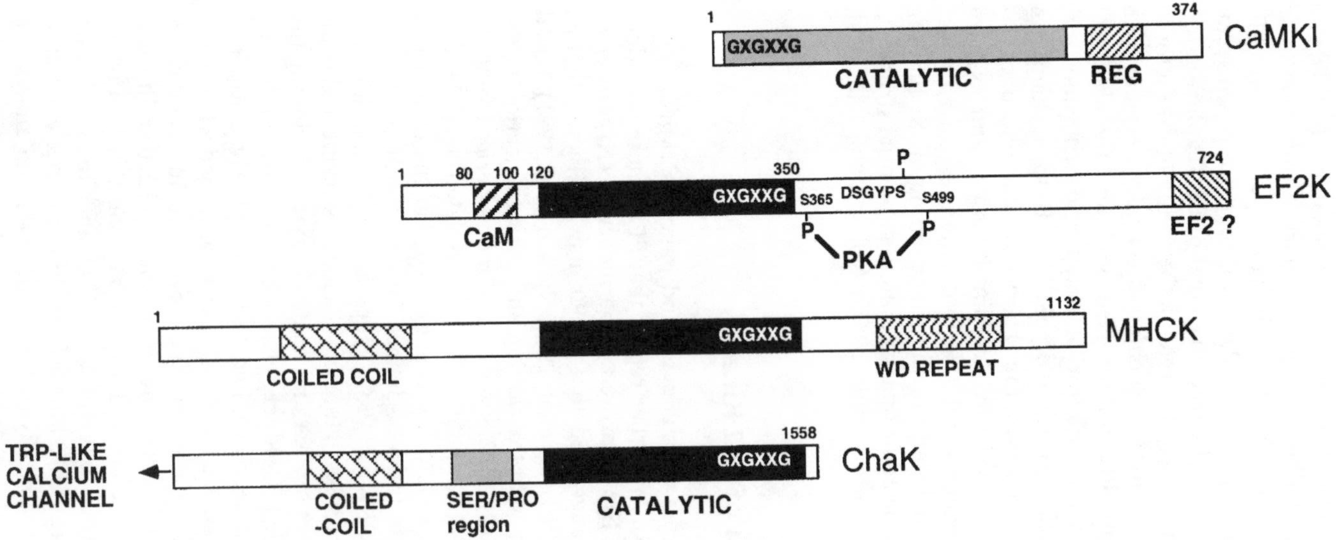

Fig. 2. Domain organization of a novel family of protein kinases. The illustration shows a comparison of CaM kinase I (*CaMKI*), EF2 kinase (*EF2K*), myosin heavy chain kinase (*MHCK*), and the C-terminal part of *ChaK*, which contains the conserved EF2 kinase/MHCK catalytic domain. CaM kinase I contains a catalytic domain related to the PKA family of protein kinases; a GXGXXG motif that is involved in MgATP binding, and a C-terminal regulatory domain that is involved in autoinhibition and CaM binding. EF2K, MHCK, and ChaK contain a unique catalytic domain; a GXGXXG motif at the C-terminal end of the conserved catalytic core whose function is not known. In EF2K, the CaM-binding domain is located N-terminal to the catalytic core; phosphorylation by PKA at Ser365 and Ser499, and phosphorylation of Ser444 within a DSGYPS motif may be involved in regulated proteolysis of the enzyme, the extreme C-terminus of the protein may be involved in binding eEF2

distinct from either EF2 kinase or MHCK (Fig. 3). Deletion of this domain in MHCK or EF2 kinase supports the contention that it comprises the catalytic core of both enzymes (see below). In addition, we have characterized one of the novel gene sequences (termed ChaK) in more detail and have isolated a full-length murine cDNA. Expression of recombinant fragments of ChaK in *E. coli* has confirmed that it is a protein kinase; the enzyme autophosphorylates on serine, and binds the ATP analog, FSBA (p-fluorosulphonylbenzoyl 5′-adenosine). However, ChaK does not phosphorylate eEF2 or a peptide derived from the phosphorylation site of *Dictyostelium* myosin heavy chain. We have recently completed sequencing the full-length ChaK cDNA and it appears to be a unique protein. The conserved EF2 kinase-related domain is at the extreme C-terminus of a large protein of 1863 amino acids. Notably, the N-terminal part of ChaK (for Channel Kinase) is related to the *Drosophila* Trp and capsaicin family of Ca^{2+} channels (Harteneck et al. 2000).

The identification of ChaK as a protein kinase reinforces the idea that EF2 kinase is a member of a growing family of protein kinases distinct from the prototypical PKA enzyme family. Ryazanov and colleagues have suggested that this new family be termed "alpha kinases" since they speculate that these enzymes recognize target phosphorylation sites located in α-helices within their respective substrates. Whether this is true of EF2 kinase is unclear, however, comparison of eEF2 with the known structures of EF-Tu and EF-G places Thr56 and Thr58 in eEF2 within the conserved E domain that is highly flexible and is not predominantly α-helical (Fig. 1).

The sequence of the catalytic domain of the EF2 kinase family reveals perhaps as many as ten subdomains, each of which contains a variety of identical amino acids (Fig. 3). Within these subdomains there are no obvious residues that would correspond to the critical substrate binding and catalytic amino acids found in the PKA family of protein kinases. A GXGXXG motif towards the C-terminus of the catalytic domain of the EF2 kinase family might correspond to a similar motif found towards the N-terminus of the PKA catalytic domain, but this is not completely conserved in related EF2 kinase family members. Notably, mutation of either of two conserved cysteine residues (C313 and C317 in EF2 kinase) results in loss of activity (Diggle et al. 1999; M. Matsushita and A. Nairn, unpubl. results), suggesting a catalytic or structural role. The distinct catalytic domain of EF2 kinase provides an explanation for the fact that small-molecule inhibitors of the PKA family of protein kinases have little or no effect on EF2 kinase (M. Matsushita and A. Nairn, unpubl. results), whereas a novel class of selenocarbonyl compounds are specific inhibitors of EF2 kinase but not of several members of the PKA family (Cho et al. 2000).

Structure-function studies have provided additional insight into the mechanism and regulation of EF2 kinase (Diggle et al. 1999; M. Matsushita and A. Nairn, unpubl. results). Truncation mutagenesis studies indicate that the core catalytic domain of EF2 kinase, EF2 K(1–350), binds and is cross-linked to the ATP analog, FSBA. Notably, while EF2 K(1–350) can autophosphorylate itself in

```
CONSENSUS  W.......K.E..PF.E..GA.R.A..T...........................................Y..K.............
M.ChaK     WSQLGLCAKIEFLSKEEMGGGLRRAVKVLCTWSE-----------------------------HDILLSGHLYIIKSFLPEVINIWSSIY
M.EF2K     WLKDEVLIKMASQPFGR--GAMRECFRTKKD------------------------SNFLHAQ----QWKGASNYVAKRYLE----------
D.MHCKA    WIRLSMKLKVERKPFAE--GALREAYHTVSLGVGTDENYPLGTTTKLFPPIEMISPISKNNEAMTQLKNGTKFVLKLYKKEA--------
D.MHCKB    WTCTATLVKVEPVPFAE--GAFRKAYHTLDL----------------------------SKSGASGRYVSKIGKK----------
D.MHCKC    WTHSIVCVSIEKTPFAK--GSCRTAHKLKDW-----------------------------SQPDQGLVGKFSTNK----------
HEART K    GDHLRGQISTEELHFGE--GVHRKAFRSKVMQGLM----------------------------PVFQPGHACVLKVHNAVAHGTRNNDE
CH 4K      AQETIVYLGDYLTVKKK--GRQRNAFWVHHLHQEEIL--------------------------------GRYVGKDYKEQ---------
           ___________________________________                                      __________
                            I                                                               II

CONSENSUS  .....R..YFEDV.MQM.AKKW..KFN..KPPK..I.FL.S.V.EL.DR.........C..EP..EG.Y.KYNNNSG.V..D......
M.ChaK     KEDTVLHLCLREIQQQRAAQKLTFAFNQMKPKSL---PYSPRFLEVFLLYCHSAGQ-WF-AVEECMTGEFRKYNNNNGDEIIPTNT---
M.EF2K     --PVDRSVYFEDVQLQMEAKLWGEDYNRHKPPKQ-VDIMQMCIIELKDRPGQP----LF-HLEHYIEGKYIKYNSNSGFVRDDN-----
D.MHCKA    EQQASRELYFEDVKMQMVCRDWGNKFNQKKPPKK-IEFLMSWVVELIDRSPSSNGQPILCSIEPLLVGEFKKNNSNYGAVLTN------
D.MHCKB    --PTPRPSYFEDVKMQMIAKKWADKYNSFKPPKKKIEFLQSCVLEFVDRTSSD----LICGAEPYVEGQYRKYNNNSGFVSNDE-----
D.MHCKC    -KKTTRDSYFTDVLMQTFCAKWAEKFNEAKPPKP-ITFLPSYVYELIDHPPPY----PVCGGEPFIEGDYKKHNNNSGYVSSDA-----
HEART K    LVQRNYKLAAQECYVQNTARYYAKIYAAEAQPLEGFGEVPEIIPIFLIHRPENNIPYATV--EEELIGEFVKYSIRDGKEINFLRRDSE
CH 4K      ---KGLWHHFTDVERQMTAQHYVTEFNKRLYEQNIPTQIFYIPSTILLILEDKTIKGCISV-EPYILGEFVKLSNNTKVVKTEYKATEY
              _______________________________     _________________         ___________________
                            III                           IV                           V

CONSENSUS  .R.TPQAFSHFTYE.SN.QLLVVDIQGVGD.......YTDPQIHT.......G.GFG.GNLG..G..KF..THKCN..C..L.L....
M.ChaK     NEEIMLAFSHWTYEYTRGELLVLDLQGVGE-------NLTDPSVIKAEEKTSCDMVFGPANLGEDAIKNFRAKHHCNSCCRKLKL----
M.EF2K     IRSTPQAFSHFTFERSGHQLIVVDIQGVGD-------LYTDPQIHTEK-----GTDFGDGNLGVRGMALFFYSHACNRICQSMGL----
D.MHCKA    -RSTPQAFSHFTYELSNKQMIVVDIQGVDD-------LYTDPQIHTPD-----GKGFGLGNLGKAGINKFITTHKCNAVCALLDL----
D.MHCKB    -RNTPQSFSHFTYEHSNHQLLIIDIQGVGD-------HYTDPQIHTYD-----GVGFGIGNLGQKGFEKFLDTHKCNAICQYLNL----
D.MHCKC    -RNTPQSFSHFSYELSNHELLIVDIQGVND-------FYTDPQIHTKS-----GEGFGEGNLGETGFHKFLQTHKCNPVCDFLKL----
HEART K    AGQKCCTFQHWVYQKTSGCLLVTDMQGVGM-------KLTDVGIATLA---RGYKGFK-GNCSMTFIDQFRALHQCNKYCKMLGLKSL-
CH 4K      ----GLAYGHFSYEFSNHRDVVVDLQGWVTGNGKGLIYLTDPQIHSVD-----QKVFTTNFGKRGIFYFFNNQHVECNEICHRLSLTRP
             ______________    ____________       _________      _________     __________________
                   VI              VII               VIII            IX                 X
```

a Ca^{2+}/CaM-dependent manner, this truncated protein does not phosphorylate eEF2 (Fig. 4). Indeed, removal of the C-terminal 34 amino acids of the enzyme EF2 K(1–690) results in an almost total loss of EF2 kinase activity, but not of autophosphorylation, suggesting that a binding site for eEF2 may be located at the extreme C-terminus of the kinase. Comparison of EF2 kinase from different species, including mammals and *C. elegans* indicates that a stretch of about 100 amino acids at the C-terminus of the protein is highly conserved (Ryazanov et al. 1997). Addition of a recombinant fragment including residues 589–724 inhibits eEF2 phosphorylation, supporting the idea that this C-terminal region may be involved in EF2 binding, and perhaps orienting the protein such that Thr56/Thr58 interacts with the catalytic domain (M. Matsushita and A. Nairn, in prep.). This conclusion is consistent with the observation that EF2 kinase fails to phosphorylate short peptides that encompass Thr56/Thr58 (Redpath et al. 1993). In contrast to EF2 kinase, the catalytic domain of MHCK is active towards a short synthetic peptide derived from the phosphorylation site of myosin heavy chain (Côté et al. 1997). However, the WD repeat domain that follows the catalytic domain of MHCK binds to myosin heavy chain and significantly increases its rate of phosphorylation (Kolman and Egelhoff 1997), suggesting that the eEF2 kinase/MHCK family may interact with their respective peptide substrates at regions outside of the catalytic domain.

EF2 kinase is essentially inactive in the absence, but highly active in the presence, of Ca^{2+}/CaM. In other Ca^{2+}/CaM-dependent protein kinases like CaM kinase I (Goldberg et al. 1996), CaM binds to a regulatory domain that follows the catalytic domain (see Fig. 2). In the basal state, CaM kinase I is maintained in an inactive autoinhibited state via a pseudosubstrate sequence in the regulatory domain that interacts with amino acids around and within the active site of the kinase. Binding of CaM to the regulatory domain of CaM kinase I relieves these autoinhibitory interactions and allows access of MgATP and peptide substrate to the active site. Site-directed mutagenesis, as well as peptide binding and competition studies, have localized the CaM-binding domain to residues 80–100 of EF2 kinase (Fig. 5). Thus, the domain organization of the regulatory regions of eEF2 kinase is also different from that of other Ca^{2+}/CaM-dependent protein kinases.

In summary, EF2 kinase is a unique member of the CaM-dependent protein kinase family that has a distinct structure and substrate specificity. The recombinant full length protein efficiently phosphorylates eEF2 in a Ca^{2+}/CaM-

Fig. 3. Comparison of the amino acid sequences of the catalytic domains of EF2 kinase and related kinases. The amino acid sequences of the conserved catalytic domain of ChaK, EF2 kinase (EF2K), three isoforms of MHCK from dictyostelium, and two unique but related kinase domains identified in EST databases, are illustrated. Up to ten distinct subdomains (*I-X*) can be identified that contain amino acids that are completely conserved (*bold letters*) or partially conserved (see consensus sequence). The close relationship between EF2K, MHCK, and some of the other sequences has also been recognized by Egelhoff and Ryazanov. (Ryazanov et al. 1997, 1999)

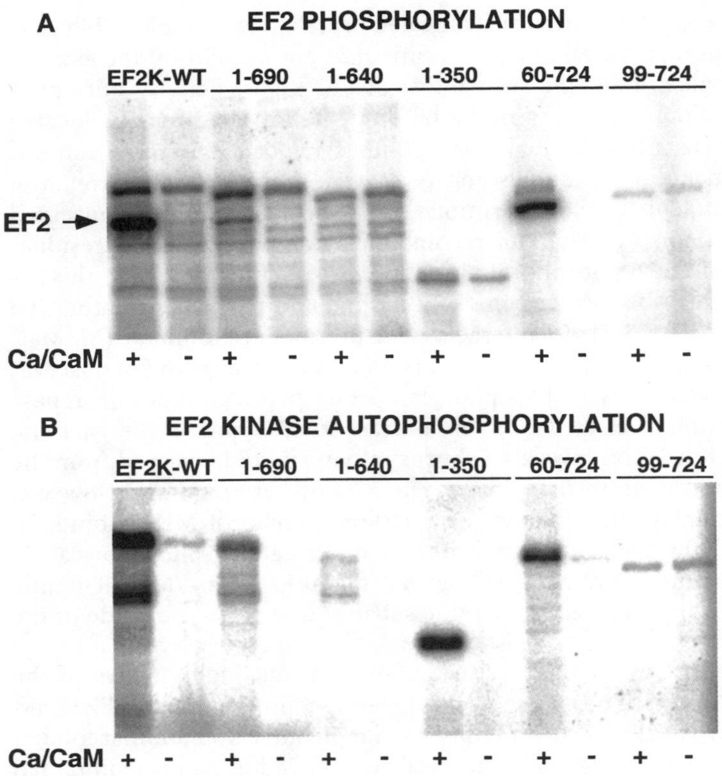

Fig. 4A,B. Autophosphorylation and phosphorylation of EF2 by truncation mutants of EF2 kinase. Wild-type EF2 kinase (EF2K-WT) and a series of C-terminal and N-terminal truncated EF2 kinase mutants were expressed as GST-fusion proteins as indicated. **A** GST-EF2K (5 μg/ml) and EF2 (20 μg/ml) were incubated with [^{32}P]ATP for 10 min at 30 °C in the absence or presence of 1.5 mM $CaCl_2$, 1 μM CaM (indicated as – or +). Proteins were separated by SDS-PAGE and autoradiography performed. The low level of incorporation of ^{32}P into GST-EF2K reflects autophosphorylation as shown more clearly in **B**. **B** EF2K-WT and EF2K mutants (10 μg/ml) were incubated with [^{32}P]ATP for 90 min at 30 °C in the absence or presence of 1.5 mM $CaCl_2$, 1 μM CaM. Proteins were separated by SDS-PAGE and autoradiography performed

dependent manner, and exhibits similar physiochemical properties to the enzyme purified from rat pancreas or rabbit reticulocytes. The organization of EF2 kinase differs markedly from the PKA superfamily but its catalytic domain resembles *Dictyostelium* MHCK and other genes that are only just beginning to be characterized. Several important questions remain about the structure and regulation of eEF2 kinase and other family members. What is the structure of the catalytic domain? The distinct amino acid sequence raises the possibility of a unique three-dimensional structure. However, it should be noted that the three-dimensional structures of actin-fragmin kinase and a kinase

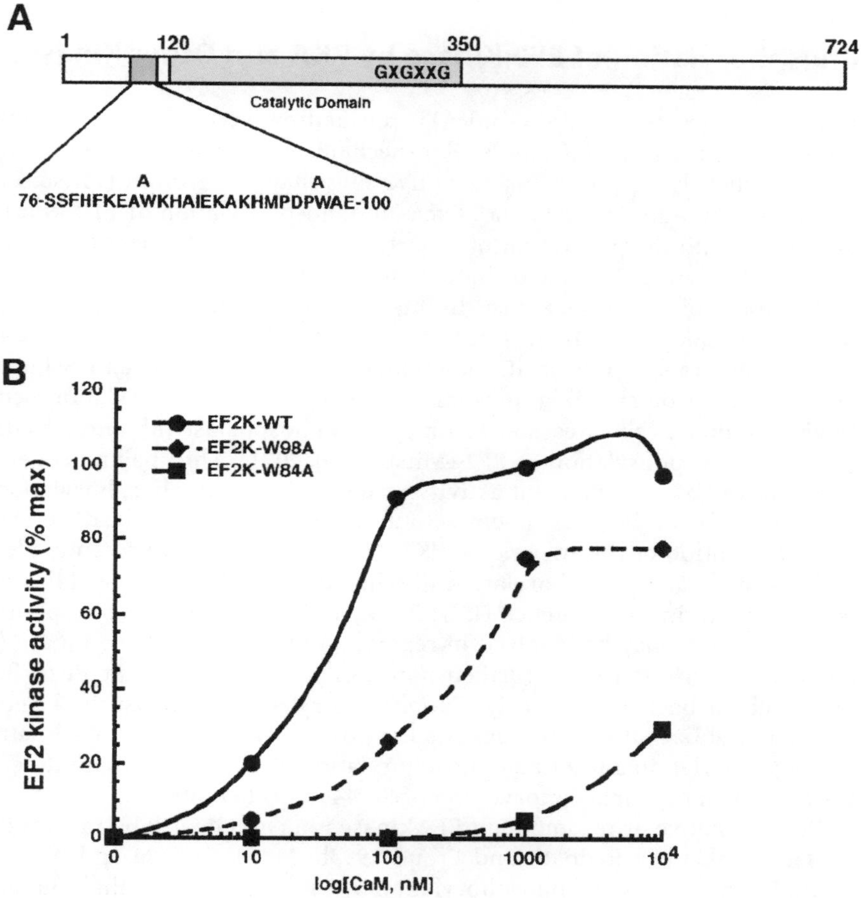

Fig. 5A, B. Identification of the calmodulin (CaM)-binding domain of EF2 kinase. **A** Schematic representation of EF2 kinase indicating the position of the CaM binding domain. **B** EF2K-WT, and mutants in which Trp84 or Trp98 were mutated to alanine (EF2K-W84 A and EF2K-W98 A) were assayed using various concentrations of CaM. These results identified residues ~80–100 as the CaM-binding domain. Peptide competition and CaM/peptide binding studies confirmed this result (not shown)

involved in aminoglycoside antibiotic resistance are very similar to that of the catalytic domain of the PKA family of proteins, despite no obvious amino acid sequence homology (Hon et al. 1997; Schlossmann et al. 2000). How is EF2 kinase maintained in an inhibited state in the absence of Ca^{2+}/CaM, and how is eEF2 recognized by the activated enzyme? Finally, as discussed further below, how is EF2 kinase activity regulated by autophosphorylation and phosphorylation by other protein kinases such as PKA?

4
Phosphorylation of EF2 Kinase by PKA and Other Kinases

In the presence of Ca^{2+}/CaM and ATP, purified native EF2 kinase autophosphorylates through an intramolecular mechanism. Both serine and threonine are phosphorylated, and as many as five sites may be involved (Mitsui et al. 1993; Redpath and Proud 1993a). Such autophosphorylation of EF2 kinase is associated with the generation of a partially Ca^{2+}/CaM-independent enzyme activity, suggesting that one or more autophosphorylation sites may interfere with autoinhibition of the kinase. In this respect, autophosphorylation resembles that of CaM kinase II, in which case generation of Ca^{2+}/CaM-independent activity plays a key role in its functional properties (Braun and Schulman 1995). A major outstanding question is whether autophosphorylation occurs in vivo. In intact cells, treatment with protein phosphatase inhibitors leads to increased phosphorylation of EF2 (Mitsui et al. 2001). This may reflect either constitutive Ca^{2+}-independent activity resulting from autophosphorylation of EF2 kinase, basal Ca^{2+}-dependent activity expressed at resting Ca^{2+} levels, or phosphorylation of the enzyme by PKA (see below). We have recently identified one of the autophosphorylation sites in native EF2 kinase as Ser444, which is present within the sequence DSGYP(pS), and it appears that phosphorylation of this site may be involved in regulation of turnover of the kinase (see below). In contrast to the purified native enzyme, recombinant EF2 kinase expressed in bacteria is poorly autophosphorylated (M. Matsushita and A. Nairn, unpubl. results). This suggests the possibility of a hierarchical pattern of phosphorylation and/or autophosphorylation of EF2 kinase that influences the subsequent autophosphorylation of Ser444 or other sites.

Purified native or recombinant EF2 kinase is an excellent substrate for PKA (Mitsui et al. 1993; Redpath and Proud 1993b; M. Matsushita and A. Nairn, unpubl. results). In vitro, phosphorylation by PKA also leads to the generation of significant Ca^{2+}/CaM-independent kinase activity. PKA phosphorylates EF2 kinase at multiple sites, some of which may overlap with sites that are autophosphorylated. We have recently identified Ser365 and Ser499 as the major PKA sites (M. Matsushita, A. Nairn, B. Chait, unpubl. results). The identity of these sites has been confirmed by site-directed mutagenesis. In addition, mutation of Ser499 or of Ser365 to acidic residues resulted in the generation of low Ca^{2+}/CaM-independent activity. Together, these results suggest that phosphorylation of these sites may regulate, in part, the Ca^{2+}/CaM-dependence of the enzyme. It is not immediately obvious how this occurs, given that these sites are distant from the catalytic and regulatory domains and are not conserved in EF2 kinases from different species.

The physiological consequences of phosphorylation of EF2 kinase by PKA have not been clearly established, and are likely to be complex. Brief activation of PKA in some intact cells is associated with increased eEF2 phosphorylation and increased Ca^{2+}/CaM-independent EF2 kinase activity in cell extracts

(Diggle et al. 1998; Hovland et al. 1999; Y. Shimizu and A. Nairn, unpubl. results). However, direct phosphorylation of EF2 kinase in intact cells has yet to be demonstrated. In cAMP-treated fat cells, increased eEF2 phosphorylation is associated with inhibition of protein synthesis and increased ribosomal transit time (Diggle et al. 1998). Thus, one consequence of activation of PKA may be an increase in basal EF2 kinase activity and eEF2 phosphorylation in the absence of increased intracellular Ca^{2+} levels. However, activation of PKA is frequently associated with activation of Ca^{2+} channels that may in turn lead to direct activation of EF2 kinase. In addition, longer activation of PKA is associated with downregulation of EF2 kinase levels (see below).

5
Regulation of EF2 Kinase Turnover by a Ubiquitination/Proteosome-Dependent Pathway

Using both enzymatic and immunological assays, EF2 kinase has been shown to be widely distributed in mammalian tissues with high levels being detected in pancreas and reticulocytes, and lower levels being detected in liver, kidney and muscle (Nairn et al. 1985; A. Nairn et al., unpubl. results). EF2 kinase is expressed at high levels in the brain at early stages of development, but decreases to relatively low levels in adult brain (K. Nastiuk and A. Nairn, unpubl. results). Notably, despite the conservation of the Thr56 and Thr58 phosphorylation sites, EF2 kinase is apparently absent from yeast since its DNA sequence is not found in the *S. cerevisiae* genome.

The level of expression of EF2 kinase is subject to significant regulation in a number of different cell systems. The best characterized of these is in PC12 cells, where high levels of kinase activity are downregulated over 1–2 h by treatment with nerve growth factor (NGF), insulin-like growth factor (IGF), and agents that increase cAMP levels (Nairn et al. 1987; Brady et al. 1990; H.C. Palfrey et al., in prep.). In PC12 cells, epidermal growth factor also stimulates downregulation to a small extent but the growth factor has no effect on EF2 kinase activity in A431 cells (H.C. Palfrey, unpubl. results). Forskolin stimulates downregulation in other cells of nervous system origin, as well as in non-neuronal cells, but the effect is less than that in PC12 cells (H.C. Palfrey, Shimizu, A. Nairn, in prep.). EF2 kinase is also decreased in stationary phase C6 glioma cells compared to that in proliferating cells (Bagaglio et al. 1993), and in differentiating HL-60 cells compared to proliferating cells (Nilsson and Nygård 1995). Downregulation of kinase activity has also been observed during the final stages of *Xenopus* oogenesis (Severinov et al. 1990). While EF2 phosphorylation is likely to depend principally on the level of cytoplasmic Ca^{2+}, and perhaps on the level of PP2A activity (see below), overexpression of EF2 kinase does result in elevated EF2 phosphorylation in the absence of changes in cytosolic Ca^{2+} levels (Mitsui et al. 2001). This suggests that the enzyme levels can play a role in determining the proportion of eEF2 that is phosphorylated at any given time. As discussed below, altered EF2 kinase levels may also be linked to changes in eEF2 phosphorylation observed during different stages of the cell cycle.

The original study of EF2 kinase downregulation in PC12 cells was based on enzyme activity measurements. In that study, it emerged that de novo protein synthesis was necessary for recovery of enzyme activity back to control levels following removal of the stimulus (NGF or forskolin; Nairn et al. 1987). This suggested that downregulation might involve degradation of EF2 kinase rather than a reversible modification of enzyme activity. This idea has been confirmed by recent studies using an antibody specific for EF2 kinase (H.C. Palfrey et al., in prep.). Using "wild-type" PC12 cells, or cells expressing tenfold higher levels of TrkA receptors, we have recently found that both NGF and forskolin lead to the rapid loss of immunodetectable EF2 kinase. Importantly, the cAMP- and NGF-regulated downregulation of EF2 kinase is blocked by lactacystin, a fairly specific inhibitor of ubiquitin/proteosome-dependent proteolysis. Recent studies in other systems have revealed that the inhibitors of NFκB (IκB) (Chen et al. 1996; Yaron et al. 1997) and β-catenin (Aberle et al. 1997) are subject to phosphorylation-dependent targeting to the ubiquitination-proteosome pathway (Hershko 1997; Hershko and Ciechanover 1998; Yaron et al. 1999). This process is critical for the regulation of NFκB-dependent transcription, and for Wnt-dependent signaling. Both IκB and β-catenin contain a conserved motif (DSGhxS) which includes two serine residues that are critical for their phosphorylation-dependent degradation. We have identified a related motif in EF2 kinase that includes Ser444, the site that is apparently autophosphorylated. The two conserved serine residues in IκB are phosphorylated by IκB kinase, an enzyme regulated by a complex signaling pathway downstream of the tumor necrosis factor (TNF) receptor (Scheidereit 1998). In contrast, the two conserved serine residues in β-catenin are phosphorylated by GSK-3β, an enzyme regulated by the Wnt signaling pathway (Aberle et al. 1997).

PKA is clearly involved in the downregulation of EF2 kinase in cells exposed to agents that elevate cAMP levels. PKA may also be involved in the downregulation response to NGF since PC12 cells deficient in PKA downregulate EF2 kinase poorly in response to NGF (Brady et al. 1990). Other studies indicate that p21ras plays a role in linking activation of TrkA NGF receptors to EF2 kinase, but the connection between this signaling pathway and PKA is not clear. EF2 kinase is a substrate for PKA in vitro, and the most obvious model is that direct phosphorylation of EF2 kinase by PKA triggers downregulation. Ser365 and Ser499, the sites in EF2 kinase phosphorylated by PKA, flank the DSGhxS motif, and may influence autophosphorylation of Ser444, or influence the ability of Ser444, or Ser440 (the preceding serine residue), to be phosphorylated by one or more kinases. Notably, EF2 kinase is also phosphorylated in vitro by GSK-3β although the sites remain to be identified (M. Matsushita and A. Nairn, unpubl. results). In any event, these observations raise the likelihood that the degradation of IκB, β-catenin and EF2 kinase occur via a related, though distinct, mechanism that involves phosphorylation-dependent targeting to the ubiquitination-proteosome pathway.

6 Hormonal Control of eEF2 Expression and Phosphorylation

Cell growth and differentiation are regulated by an array of overlapping signaling pathways, and growing evidence indicates that translational control plays an essential role in the response of cells to mitogenic signals (Redpath and Proud 1994; Sonenberg 1996). Phosphorylation of several initiation factors is associated with increased cell growth. For example, initiation factor eIF4E phosphorylation is increased in ras-transformed fibroblasts (Frederickson et al. 1992; Rinker-Schaeffer et al. 1992; Sonenberg and Gingras 1998), and overexpression of this factor leads to transformation of fibroblasts (Lazaris-Karatzas et al. 1990). In contrast, phosphorylation of initiation factor eIF2α by a heme-regulated protein kinase, or by a double stranded RNA-dependent protein kinase (PKR), and other kinases related to PKR, inhibits protein synthesis (Rhoads 1993; Redpath and Proud 1994; Clemens 1996; Shi et al. 1998; Harding et al. 2000), a process that is important in serum-deprived cells, and in heat shock and viral infection (Clemens 1996; Taylor et al. 1999).

A critical pathway in the regulation of translation by growth factors involves the activation of the mammalian target of rapamycin, mTOR (Jefferies and Thomas 1996; Brunn et al. 1997; Jefferies et al. 1997; Avruch 1998; Gingras et al. 1998). Rapamycin is a bacterial macrolide that acts as an immunosuppressant by blocking progression through the G1 phase of the cell cycle. While many details remain to be clarified, various members of the rapamycin-sensitive pathway have been identified, and the proposed role of some of these are illustrated in Fig. 6. Receptor tyrosine kinases (RTK) are linked via activation of PI3 kinase (PI3K) to activation of protein kinase B (PKB, or akt), and this part of the pathway is sensitive to the PI3K inhibitor, wortmannin. PKB most likely directly activates mTOR (or FRAP, FKBP12-rapamycin-associated protein), at which point the pathway bifurcates. mTOR activates either directly or indirectly, p70 S6 kinase (p70S6K), which in turn phosphorylates ribosomal protein S6. mTOR also regulates the availability of initiation factor eIF4E through the phosphorylation of the eIF4E-binding protein, PHAS-1 (or 4E-BP) (Gingras et al. 1999), and leads to a general increase in protein synthesis. In addition, it is probable that the activation of this pathway leads to the translation of specific mRNAs, such as cyclin D, that are critical to G1 phase progression (Hashemolhosseini et al. 1998; Muise-Helmericks et al. 1998), and c-myc, an immediate early gene associated with cell growth and transformation (West et al. 1998). Finally, recent studies have indicated that the activation of the mTOR/p70S6K pathway also plays a role in the regulation of protein synthesis by amino acid and nutrient availability, via an unknown link to mTOR that does not appear to involve PI3K (Fox et al. 1998; Hara et al. 1998; Wang et al.

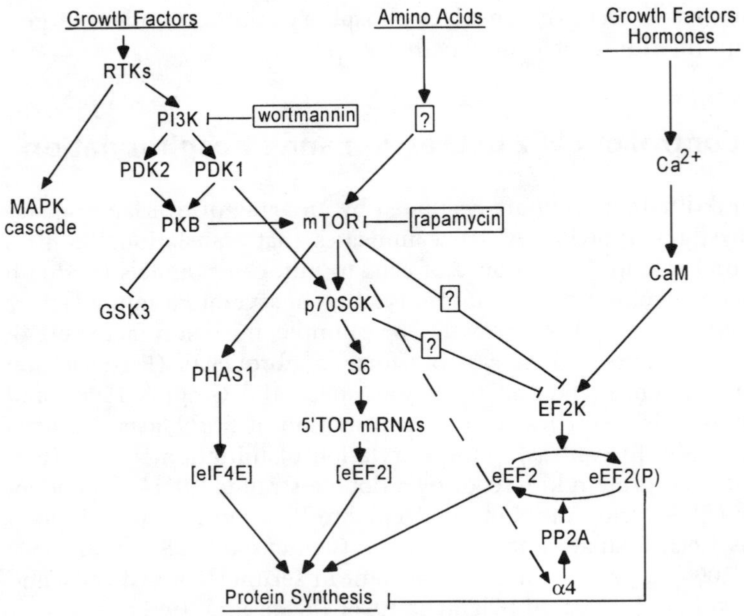

Fig. 6. Signal transduction pathways linking regulation of protein synthesis to growth factor receptors. Recent studies have established a complex array of signal transduction pathways linking extracellular factors to the regulation of protein synthesis (Jefferies and Thomas 1996; Abraham 1998; Avruch 1998). Some, but not all, of the potential pathways are illustrated. For example, PKC isoforms are also downstream of PI3 kinase (Chou et al. 1998; Le Good et al. 1998) and are likely to regulate protein synthesis (Mendez et al. 1997). Protein synthesis initiation is regulated through the availability of eIF4E. Specific synthesis of 5′TOP mRNAs for elongation factors, e.g., eEF2, and ribosomal proteins, is regulated by phosphorylation of S6 by p70S6 kinase. eEF2 dephosphorylation is also regulated by an interaction of the α4 protein with the catalytic subunit of PP2A. α4 is regulated by the rapamycin-sensitive pathway, possibly by direct phosphorylation by mTOR (Jiang and Broach 1999; *dashed line*), although this is yet to be demonstrated in mammalian cells. *Arrowheads* designate an activating relationship; *flat arrows* designate an inactivating relationship

1998; Xu et al. 1998). The mTOR pathway also potentially impinges on the EF2 kinase-eEF2 system at various points, as summarized below.

6.1 Regulation of eEF2 Protein

eEF2 levels in most cells seem to be approximately stoichiometric with the number of ribosomes (Gill and Dinius 1973; Nygård and Nilsson 1984). Given this parity, an alteration in eEF2 protein levels is likely to influence the overall rate of protein translation. Several studies indicated that eEF2 biosynthesis is subject to regulation (Levenson et al. 1989; Rattan 1991; Vary et al. 1994). In particular, insulin rapidly increases the synthesis of eEF2 prior to the general increase in protein synthesis that takes place later in response to the hormone. The molecular basis for this action of insulin has been largely elucidated and mTOR is known to play a prominent role (Fig. 6). Translation of a class of mRNAs that contain a polypyrimidine tract at their 5′ end (5′TOP mRNAs), notably, mRNAs encoding eEF2, eEF1α, and various ribosomal proteins, are coordinately regulated by a rapamycin-sensitive signaling pathway (Jefferies et al. 1994; Terada et al. 1994, 1995). Of these various proteins, eEF2 represents a major target for regulation by rapamycin. As mentioned above, this most likely occurs through the regulation of p70S6K, the consequent increase in phosphorylation of ribosomal protein S6, and the specific recruitment of 5′TOP mRNAs to the ribosome (Fig. 6), thus coupling mitogenic signals to the biogenesis of the translational machinery.

6.2 Regulation of eEF2 Phosphorylation

Besides modulation of eEF2 protein levels, the rapamycin-sensitive pathway may also influence eEF2 phosphorylation. Rapamycin treatment in COS-7 cells resulted in increased phosphorylation of eEF2 (Chung et al. 1999). Proud and coworkers showed, using CHO cells which overexpress the insulin receptor, that insulin treatment resulted in dephosphorylation of eEF2, and that this was prevented by either rapamycin or wortmannin (Redpath et al. 1996b). Similar results have been observed in rat cardiac myocytes (Wang et al. 2000) and in rat adipocytes and 3T3-L1 cells (Diggle et al. 1998). A striking feature of these studies is the unusually high basal level of phospho-eEF2, most likely due to the extended serum starvation protocol employed, since we have found that serum removal in various cell types leads to a substantial increase in phospho-eEF2 that is reversed by serum re-addition (K. Nastiuk et al., unpubl. results). Proud and coworkers have also shown that the effect of insulin in CHO cells requires the presence of both glucose and amino acids (Campbell et al. 1999), consistent with the presence of convergent signaling pathways that couple the actions of insulin to the nutrient status of the cell.

An important outstanding question is the mechanism by which eEF2 phosphorylation is regulated by the rapamycin-dependent signaling pathway. One possibility is that EF2 kinase activity could be inhibited following phosphorylation by one of the upstream kinases in the pathway (see Fig. 6; mTOR and p70S6K are illustrated as potential EF2 kinase kinases, but PKB is also a viable candidate). Treatment with insulin results in a rapid decrease in EF2 kinase activity in fat and heart cell extracts (Redpath et al. 1996b; Diggle et al. 1998; Wang et al. 2000), suggestive of a modification of enzyme activity as opposed to degradation. (It is also possible that EF2 phosphatase activity could be affected by insulin; see below.) In vitro, p70S6K phosphorylates EF2 kinase but has no effect on activity (Redpath et al. 1996b). An alternative possibility is that the EF2 kinase protein level, but not necessarily its activity, is downregulated by phosphorylation of EF2 kinase by some component of the rapamycin-sensitive signaling pathway. This is an attractive hypothesis given the discovery that EF2 kinase may be subject to phosphorylation-dependent targeting of the ubiquitination/proteosome pathway (see above). In support of this possibility, preliminary results have shown that the level of EF2 kinase protein increases approximately threefold upon serum deprivation from growing and dividing cells, and that insulin treatment rapidly reduces EF2 kinase levels following addition to serum-deprived cells (Y. Shimizu et al., unpubl. results). However, in PC12 cells, the effect of NGF is not apparently prevented by wortmanin or rapamycin, raising the possibility of multiple pathways being involved in regulation of EF2 kinase levels.

6.3 Regulation of eEF2 Dephosphorylation

Following various physiological stimuli, the phosphorylation of eEF2 is transient, reflecting the rapid activation of EF2 kinase activity by transient increases in Ca^{2+} concentration followed by eEF2 phosphatase activity (Palfrey et al. 1987; Hincke and Nairn 1992; Marin et al. 1997; Scheetz et al. 1997). The latter is probably PP2A, based on in vitro biochemical studies and the effect of specific phosphatase inhibitors such as calyculin A in vivo (Nairn and Palfrey 1987; Redpath and Proud 1989, 1990).

In addition to the regulation of EF2 kinase, eEF2 dephosphorylation by PP2A might also be modulated by rapamycin-sensitive signaling pathways. Recent studies have provided support for the idea that regulation of PP2A, via a novel mechanism, is involved in the dephosphorylation of eEF2 (Chung et al. 1999). This work followed from the observation that a protein, termed α4, associates with the catalytic subunit of PP2A (and related PP2A catalytic subunits) (Murata et al. 1997; Chen et al. 1998). α4 is a mammalian cell homologue of the Tap42 protein of *S. cerevisiae* that, in yeast, is a component of the TOR signaling pathway (Di Como and Arndt 1996) and may be a direct substrate for TOR (Jiang and Broach 1999). Notably, Tap42 interacts with the catalytic subunit of PP2A and is regulated by nutrient growth signals and the rapamycin-sensitive TOR pathway (Jiang and Broach 1999). In mammalian cells, mutations in regulatory sites of the PP2A catalytic subunit led to its preferential interaction with α4 (Chung et al. 1999). In addition, transient expression of α4 in COS-7 cells resulted in reduced phosphorylation of eEF2, but had no effect on phosphorylation of p70S6K or PHAS-1. Similar results were obtained in either the absence or presence of rapamycin. Intriguingly, eEF2 and α4 have been recently identified, using a differential display technique, to be highly expressed in prolactin-independent rat lymphoid cells (Too 1997), raising the possibility that these two proteins are involved in the transition to growth factor independence in lymphoid tumors. Together, these results suggest that α4 functions as a novel PP2A subunit that promotes eEF2 dephosphorylation, and that the PP2A/α4 complex may function downstream of mTOR. Since eEF2 is dephosphorylated by PP2A, the simplest explanation would be that PP2A/α4 dephosphorylates eEF2. However, it is also possible that EF2 kinase could be an alternative, or additional, target for PP2A/α4. Dephosphorylation of EF2 kinase at a key regulatory site might inactivate the enzyme, or influence its degradation. Alternatively, altered PP2A activity might influence some phosphorylated component of the ubiquitination-proteosome pathway.

In addition to rapamycin-sensitive pathways, other mechanisms may also be involved in regulation of eEF2 dephosphorylation. In cardiac myocytes, angiotensin II treatment decreases eEF2 phosphorylation (A. Everett et al. 2001). The effect is blocked by okadaic acid and fostriecin (inhibitors of PP2A) but not by FK506 (an inhibitor of PP2B). Angiotensin II also activated the MEK/MAP kinase pathway, and blockade of MAP kinase (MAPK) activation with PD98059 inhibited eEF2 dephosphorylation. PI3K signaling upstream of mTOR also appears to play a role in angiotensin II signaling since eEF2 dephosphorylation could be blocked with LY29004 but not by rapamycin.

7
Relationship of eEF2 Phosphorylation to Protein Synthesis and Growth in Proliferating Cells

Many studies have indicated that the major factor that influences eEF2 phosphorylation is the level of intracellular Ca^{2+} (Palfrey et al. 1987; Mackie et al. 1989; Hincke and Nairn 1992; Marin et al. 1997; Scheetz et al. 2000; for review see Proud 1992). Through the use of peptide mapping, isoelectric focusing, and immunoblotting with a phospho-Thr56 antibody, it appears that Thr56 is the major site phosphorylated in response to activation of EF2 kinase. Increased eEF2 phosphorylation appears to be independent of the mechanism(s) involved in elevating intracellular Ca^{2+}. For example, in endothelial cells, thrombin or histamine increase eEF2 phosphorylation presumably via stimulation of release of Ca^{2+} from IP3-sensitive stores (Mackie et al. 1989). By contrast, in neurons, glutamate increases eEF2 via entry of Ca^{2+} through channels in the plasma membrane (Marin et al. 1997; see below). In many cases, the increased eEF2 phosphorylation is transient and parallels the intracellular Ca^{2+} transient produced by the specific ligand used, with eEF2 dephosphorylation occurring rapidly as Ca^{2+} levels decrease back to baseline (Palfrey et al. 1987). Although transient, the phosphorylation of eEF2 can reach high levels in intact cells stimulated with physiological agonists. For example, in endothelial cells the proportion of phospho-EF2 rises from about 15% in the basal state to about 90% after incubation with thrombin for 2 min (Mackie et al. 1989). In cultured neurons, eEF2 phosphorylation is very low but about 44% is phosphorylated after incubation with glutamate for 2 min (Marin et al. 1997; see below). Moreover, in neurons, immunocytochemical analysis indicated that eEF2 phosphorylation was increased in only about 50% of the neurons while no detectable increase was observed in the other cells, suggesting that in responsive cells eEF2 phosphorylation was virtually stoichiometric. Finally, in intact reticulocytes, incubation with a Ca^{2+} ionophore results in approximately 50% of eEF2 being phosphorylated within 2 min (Mitsui et al. 2001). Thus, phosphorylation of eEF2 is a robust response to changes in intracellular Ca^{2+}, is ubiquitous in eukaryotic cells, and is likely to play an important role in Ca^{2+}-dependent inhibition of protein synthesis in many cell types.

As phosphorylated eEF2 is unable to participate in mRNA translocation on the ribosome, and as eEF2 is thought to be rate-limiting in this process, any reduction in the functional pool of eEF2 caused by phosphorylation of the factor should result in a corresponding reduction in protein synthesis. In addition, as discussed above, high levels of phospho-eEF2 may be able to act in a dominant negative manner. Several studies in intact cells have provided support for the idea that phosphorylation of eEF2 is associated with inhibition of the elongation step in polypeptide translation. The phosphorylation of eEF2 is high in serum-deprived CHO-T cells. Following treatment with insulin, eEF2 phosphorylation is significantly decreased, and this is associated with a significant decrease in ribosomal transit time (Redpath et al. 1996b). In addition, stimulation of adipocytes with agents that increase PKA activity, is associated with increased eEF2 phosphorylation and about a three- to fourfold increase in ribosomal transit time (Diggle et al. 1998). In another study using endothelial cells, various physiological and pharmacological treatments were used to either stimulate EF2 kinase or inhibit PP2A (Mitsui et al. 2001). The phosphorylation of eEF2 was inversely proportional to the rate of protein synthesis under several different conditions. Moreover, treatment with the Ca^{2+} ionophore, ionomycin, caused a two- to threefold decrease in elongation rate, as measured by the determination of ribosomal transit times. In these latter studies, there was a good correlation between the percent inhibition of protein synthesis and the increase in ribosomal transit time, obtained using ionomycin or the elongation inhibitor, cycloheximide. In intact reticulocytes treated with a Ca^{2+} ionophore, eEF2 is rapidly phosphorylated and protein synthesis is rapidly inhibited (Mitsui et al. 2001), an effect that is correlated with elongation block and an increase in polysomes (Wong et al. 1991). Finally, in studies of neurons in culture, there was a very close inverse correlation between eEF2 phosphorylation and the rate of protein synthesis, following treatment with glutamate and other agonists that raise intracellular Ca^{2+} (see Fig. 8).

In certain cases, it has been argued that transient elevation of intracellular Ca^{2+} does not alter the rate of protein synthesis. For example, in GH_3 cells, no effect of Ca^{2+} mobilizing agents on elongation rate could be observed (Laitusis et al. 1998). The reasons for this result are not clear but may reflect the apparently high basal level of eEF2 phosphorylation measured in these cells, and the small increases in eEF2 phosphorylation obtained using the particular Ca^{2+} mobilizing agents. Under certain conditions, Ca^{2+} may also inhibit protein synthesis at the level of initiation via activation of double-stranded RNA-stimulated protein kinase (PKR) (or possibly other kinases related to

PKR such as PERK) and phosphorylation of eIF2α (for extensive review of this literature, see Palfrey and Nairn 1995; Nairn and Palfrey 1996). While the precise mechanism by which PKR is regulated by Ca^{2+} is not clear, the phosphorylation of eIF2α by PKR is apparently linked to depletion of Ca^{2+} from intracellular stores that may be linked to stress responses in the endoplasmic reticulum. In a number of cell types that have been analyzed, eEF2 phosphorylation appears to occur more rapidly than eIF2α, and to be associated with changes in Ca^{2+} levels that are under more physiological control. In CHO and HEK293 cells, brief treatment with ionomycin leads to large increases in eEF2 phosphorylation without any change in eIF2α phosphorylation (Mitsui et al. 2001). In neurons, glutamate treatment leads to Ca^{2+} influx through plasma membrane channels and results only in changes in eEF2 phosphorylation. In addition, treatment of neurons with zinc or hydrogen peroxide leads to rapid phosphorylation of eEF2 that is associated with an initial phase of inhibition of protein synthesis, and to a slower phosphorylation of eIF2α that is also associated with a more prolonged inhibition of protein synthesis (Alirezaei et al. 1999, 2001). Finally, overexpression of EF2 kinase leads to increased phosphorylation of eEF2 and inhibition of protein synthesis, while no changes in eIF2α phosphorylation are observed (Mitsui et al. 2001). Thus, phosphorylation of eEF2 and inhibition of elongation are likely to be responsible, at least in part, for the inhibition of protein synthesis observed at micromolar intracellular Ca^{2+} levels in a variety of cell types.

8
Cell Cycle-Dependent Phosphorylation of eEF2

Ca^{2+} is widely recognized as an essential messenger in eukaryotic systems regulating many intracellular processes, including gene expression and cell growth and proliferation (Berridge 1994, 1998). As a second messenger, Ca^{2+} generally acts in a transient manner, and this is also the case in cell growth and division. In *Xenopus* oocytes, fertilization is accompanied by an explosive Ca^{2+} wave that is required for the exit from meiosis (Whitaker and Patel 1990; Lu and Means 1993). Increases in cell Ca^{2+} levels have also been measured at transition points in the cell cycle (Poenie et al. 1985, 1986; Groigno and Whitaker 1998). In addition, many growth factors and mitogens cause a rapid and transient increase in intracellular Ca^{2+} levels, which is associated with a stimulation of entry of quiescent cells into the G_1 phase of the cell cycle. It is unclear if in all cases Ca^{2+} is required for the mitogenic actions of these agents (Jaye et al. 1992; Peters et al. 1992), however, it is likely that Ca^{2+} plays at least an important modulatory role in the regulation of transcription and translation that occurs upon stimulation of cell growth. In many instances, the actions of Ca^{2+} are mediated by binding to the ubiquitous binding protein CaM. Increased CaM expression occurs at the G_1/S boundary and appears to be necessary for DNA synthesis to be initiated (Means and Rasmussen 1988;

Anraku et al. 1991). CaM is activated during mitosis (Lu et al. 1993; Török et al. 1998) and is required for cell cycle progression (Anraku et al. 1991; Lu and Means 1993).

Several studies have analyzed the phosphorylation of eEF2 during the cell cycle. Phospho-eEF2 levels are elevated two- to threefold in cells synchronized by mitotic shake-off compared to cells in other stages of the cell cycle (Celis et al. 1990). We have also recently analyzed the phosphorylation of eEF2 during various stages of the cell cycle using anti-phospho-eEF2 Thr56 antibody. The phosphorylation of eEF2 was studied in several different cell types including HeLa cells, COS cells and lymphocytes, and using several different types of treatment to achieve cell synchrony (K. Nastiuk et al., in prep.). In all cases, phospho-eEF2 was high following serum deprivation and dropped upon cell cycle re-entry, subsequently increased in S phase, fell during the G2 phase, and then increased once more during M phase. These peaks of phosphorylation correspond to transition points in the levels of the G1 cyclin, cyclin E, and the G2/M cyclin, and cyclin B1 as measured by immunoblots of the same cell extracts. This complexity suggests that eEF2 phosphorylation may well play an important role in the cell cycle. For example, eEF2 phosphorylation might contribute to the marked inhibition of protein synthesis observed during mitosis (Kanki and Newport 1991; Ross 1997). Protein synthesis initiation is also inhibited during mitosis and decreased phosphorylation of eIF4E has been suggested to be responsible (Bonneau and Sonenberg 1987).

The increased phosphorylation of eEF2 during mitosis is likely to be caused at least in part by increased Ca^{2+} (Groigno and Whitaker 1998) and activation of CaM-dependent processes (Török et al. 1998). It is also possible that the level of EF2 kinase is regulated in a cell cycle-dependent fashion and an increase in the level of EF2 kinase could contribute to the phosphorylation of eEF2 observed during mitosis. In HeLa cells, EF2 kinase levels and activity are high in serum-deprived quiescent cells (K. Nastiuk et al., in prep.), consistent with the higher eEF2 phosphorylation observed at mitosis. However, in another study using Ehrlich ascites cells, EF2 kinase levels were found to be highest during the S phase (Carlberg et al. 1991). Clearly, additional studies are needed to resolve the precise mechanism(s) involved in regulation of eEF2 phosphorylation during the cell cycle.

The importance of eEF2 phosphorylation in the cell cycle is largely corroborated by recent results on the behavior of expressed eEF2 mutants in proliferating cells. To successfully probe the role of exogenous eEF2 in cells expressing high levels of the endogenous protein, we resorted to a toxin-based procedure for eliminating the latter. Notably, eEF2 is ADP-ribosylated by diphtheria and *Pseudomonas* (Pseud) A toxins resulting in the inhibition of protein synthesis and cell death. ADP-ribosylation occurs at a modified histidine residue (termed diphthamide, residue 715), and a toxin-resistant form of eEF2 was identified by Kohno and colleagues in which Gly717 was mutated to arginine (Kohno and Uchida 1987). Mouse L cells were transfected with wild-type or toxin-resistant eEF2, which retained the normal phosphorylation sites, or with toxin-resistant eEF2 in which Thr56, Thr58, or both Thr56 and 58 were mutated (Fig. 7; K. Nastiuk et al., unpubl. results). Cells were then selected by their ability to grow in the presence of normally toxic levels of Pseud A. Viable cells, therefore, were dependent on toxin-resistant eEF2 for protein synthesis and, hence, growth and proliferation. Transfection of wild-type eEF2 resulted in cells that grew normally in the absence of Pseud A but which were not viable in the presence of the toxin. Transfection of the toxin-resistant eEF2, EF2(G717R), resulted in cells that grew normally in both the presence and absence of Pseud A. Mutation of either Thr56 or Thr58 to alanine, [EF2 (G717R;T56 A) or EF2(G717R;T58 A)] resulted in viable cells which grew normally in the absence of Pseud A, but grew more slowly in the presence of the toxin. For cells expressing EF2(G717R;T58 A), the division time in the presence of toxin approximately doubled; for EF2(G717R;T56 A), the division time in the presence of toxin was approximately threefold slower. Most importantly, transfection with eEF2 mutants in which both Thr56 and Thr58 were changed to alanine [EF2(G717R;T56 A;T58 A)] resulted in no viable cells following selection in the presence of toxin.

Since no cells were obtained for the EF2(G717R;T56 A;T58 A) mutant, cells were also transfected using a plasmid conferring neomycin resistance. Subsequent selection in the presence of the antibiotic gave rise to cell lines that grew

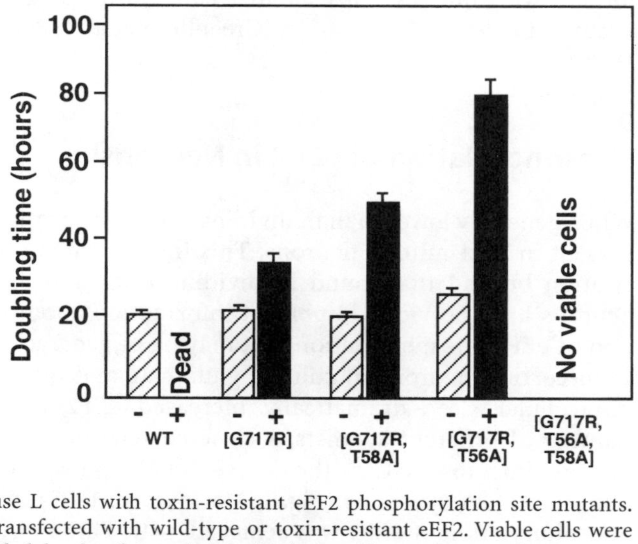

Fig. 7. Transfection of mouse L cells with toxin-resistant eEF2 phosphorylation site mutants. Mouse L cells were stably transfected with wild-type or toxin-resistant eEF2. Viable cells were selected using normally lethal levels of Pseud A toxin (50 ng/ml) over a period of several days. No cells were isolated following transfection with wild-type eEF2 or with EF2(G717;T56 A;T58 A). Cells were isolated following transfection with EF2(G717R;T56 A) or EF2(G717R;T58 A) and then grown in either the absence (−) or presence (+) of Pseud A toxin and growth rates measured by cell counting. Results from four different experiments are shown; *error bars* represent 1 SE

normally in the absence of toxin but which grew very poorly in its presence (not shown). In additional studies, cells expressing EF2(G717R;T58D) were viable but grew slowly in the presence of Pseud A, while no viable cells were obtained following transfection with EF2(G717R;T56 E) or EF2(G717R;T56 E; T58D; not shown). Cells expressing EF2(G717R;T56 S) also grew slowly in the presence of toxin, while no viable cells were obtained following transfection with EF2(G717R;T56 S;T58 A). These results suggest that the acidic residue mutants could mimic the phosphorylated state of eEF2; however, it is also possible that these results reflect the failure of the eEF2 mutants to be phosphorylated. In biochemical studies, we determined that EF2(T56 S) could be phosphorylated by EF2 kinase on serine, but this was less efficient than for the wild-type protein.

These results support a role for eEF2 phosphorylation in cell cycle progression. The difference between the growth rates (in the presence of toxin) of cells transfected with either the EF2(T56 A) or EF2(T58 A) mutants most likely reflects the relative ability of these two sites to be phosphorylated by EF2 kinase. However, the results, particularly the failure of the double mutant to grow in the presence of toxin, were unexpected. Given the observation that eEF2 phosphorylation is regulated in a cell cycle-dependent fashion, one explanation of these results is that transient inhibition of protein synthesis at the level of elongation is required for progression through a critical point in the cell cycle. For example, it has been suggested that inhibition of protein synthesis during mitosis helps stabilize labile mRNAs (Ross 1997), and this could occur at the level of elongation (Greenberg et al. 1986; Greenberg and Belasco 1993).

9
Phosphorylation of eEF2 in Neurons

While generally lower than in dividing cells, significant levels of EF2 kinase are present in post-mitotic neurons. This implies function(s) for eEF2 phosphorylation beyond that found in dividing cells, and several recent studies in neurons have provided important information about the physiological function of eEF2 phosphorylation that may be of general relevance.

In cortical neurons in culture, glutamate and other agonists that increase intracellular Ca^{2+}, dramatically increased eEF2 phosphorylation. Using a number of different agonists, a close correlation was observed between the increase in cytosolic Ca^{2+}, the degree of eEF2 phosphorylation and the inhibition of protein synthesis (Fig. 8; Marin et al. 1997). A high stoichiometry of phosphorylation of eEF2 was achieved, with Thr56 of eEF2 being phosphorylated to at least 0.4 mol/mol after a 2-min glutamate treatment. These results reinforce the conclusion from the studies described above using dividing cells that the rate of global protein synthesis is closely linked to the phosphorylation status of eEF2, which in turn is linked to cellular Ca^{2+} levels. Following

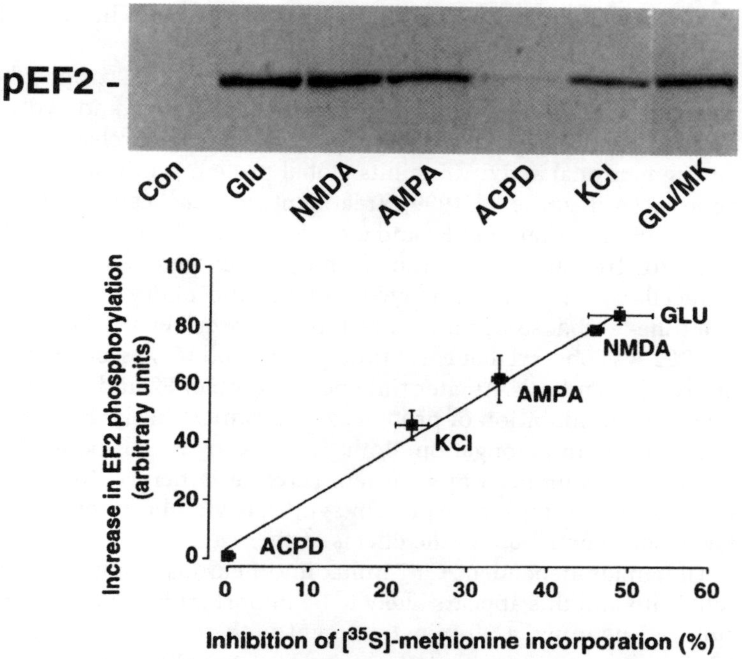

Fig. 8. Glutamate-induced phosphorylation of eEF2 and inhibition of protein synthesis in neurons. *Upper panel* Cortical neurons were exposed for 2 min to glutamate (*Glu*), NMDA, AMPA, ACPD, KCl, and glutamate plus MK-801 (*MK*), in the presence of extracellular Ca^{2+}. Samples were separated by SDS-PAGE and phospho-eEF2 (pEF2) was measured by immunoblotting. *Con* Control. *Lower panel* In separate experiments, cortical neurons were incubated with the same reagents for 5 min, during which time protein synthesis was measured by ^{35}S-methionine incorporation, and normalized to control. The relationship between eEF2 phosphorylation and inhibition of protein synthesis was then plotted. (Adapted from Marin et al. 1997)

prolonged treatment with glutamate, eEF2 phosphorylation was sustained, in contrast to other Ca^{2+}/CaM-dependent activities such as CaM kinase II. The inhibition of protein synthesis was also sustained, possibly providing an explanation for the persistent and Ca^{2+}-dependent inhibition of protein synthesis that has been described following cerebral ischemia (Orrego and Lipmann 1967; Thilmann et al. 1986; Raley-Susman and Lipton 1990). Regulation of protein synthesis in neurons following ischemia may also involve phosphorylation of initiation factor eIF2α by PKR (Hu and Wieloch 1993; DeGarcia et al. 1996). As discussed above, phosphorylation of eIF2α constitutes another potential Ca^{2+}-dependent mechanism contributing to inhibition of protein synthesis (Palfrey and Nairn 1995). However, phosphorylation of eIF2α by PKR is apparently linked to depletion of Ca^{2+} from intracellular stores (Brostrom et al. 1989; Prostko et al. 1993; Wong et al. 1993; Srivastava et al. 1995). In contrast, our results have clearly established that the Ca^{2+}-dependent regulation of eEF2 phosphorylation and inhibition of protein synthesis in cortical neurons are dependent on the presence of extracellular Ca^{2+}.

The temporal relationship between eEF2 and eIF2α phosphorylation has been analyzed in other studies in neurons using Zn^{2+} and hydrogen peroxide. Zn^{2+}, which is contained in synaptic vesicles and released with glutamate during neuronal activity, inhibits global protein synthesis in cultured cortical neurons (Alirezaei et al. 1999). Treatment with Zn^{2+} caused a transient increase in phosphorylation of eEF2 and a more sustained increase in phosphorylation of eIF2α. Treatment of neurons with hydrogen peroxide caused an increase in intracellular Ca^{2+}, and stimulated both the phosphorylation of eEF2 and eIF2α in a time- and dose-dependent fashion (Alirezaei et al. 2001). Phosphorylation of eEF2 was observed at early time points, and was associated with inhibition of protein synthesis. At later time points, both eEF2 and eIF2α were phosphorylated, and inhibition of protein synthesis most likely involved reduced rates of initiation and elongation. Both Zn^{2+} and hydrogen peroxide contribute to neuronal loss induced by transient cerebral ischemia. Together, these results suggest that inhibition of protein synthesis via phosphorylation of eEF2 and eIF2α may contribute to the effects of these agents.

Glutamate-dependent Ca^{2+} influx into neurons results in delayed neuronal cell death and this appears likely to be important in ischemia. In the studies of cultured neurons, a 30-min treatment with the glutamate receptor agonist, NMDA, induced a transient phosphorylation of eEF2, which preceded neuronal death by several hours (Marin et al. 1997). Notably, pharmacological inhibition of protein translation at the elongation stage using either cycloheximide or diphtheria toxin protected neurons against the toxicity evoked by low concentrations of NMDA. It has been proposed recently that glutamate induces either apoptotic or necrotic cell death in neurons, depending on the intensity and duration of the stimulation of NMDA receptors (Toescu 1998; Gwag et al. 1999). In certain paradigms, apoptosis requires active protein synthesis (Furukawa et al. 1997; Pike et al. 1998; Gwag et al. 1999). Therefore, phosphorylation of eEF2 and the resulting depression of protein synthesis may limit the synthesis of protein(s) required for neuronal apoptosis and, as a consequence, play an important protective role in excitotoxicity.

10
eEF2 Phosphorylation and the Regulation of Local Protein Synthesis in Neurons

Recent studies have revealed that the regulation of protein synthesis involves subcellular localization of components of the translational machinery (Hazelrigg 1998). In particular, the targeting of specific mRNA species and spatially restricted translation result in high local concentrations of protein at the site of biological function. Striking examples of mRNA localization are found in early embryonic development where specific mRNAs are localized at the animal or vegetal pole of oocytes, or at the anterior or posterior pole of young embryos. Other examples include the localization of β-actin mRNA at the leading edge of migrating fibroblasts. The mechanisms involved in subcellular targeting of mRNAs remain to be elucidated. However, specific proteins may bind to 3′ UTRs of certain mRNAs, leading to an interaction with the cytoskeleton and transport through the cytoplasm.

Growing evidence also suggests that mRNA localization plays important roles in neuronal function, including normal and abnormal neuronal development, and in synaptic plasticity (Steward 1997; Koenig and Giuditta 1999). Specific mRNAs are targeted to postsynaptic sites in dendrites, including those for glutamate receptors (both NMDA and AMPA, α-amino-3-hydroxy-5-methylisoxazole-4-propionate, subtypes) (Miyashiro et al. 1994), the α-subunit of the glycine receptor (Racca et al. 1997), the regulatory enzyme CaM kinase II (Burgin et al. 1990), structural proteins such as MAP-2 (Garner et al. 1988; Kleiman et al. 1990), and the fragile X mental retardation protein (Feng et al. 1997). Polyribosomes and other components of the translational machinery, including eEF2 (Marin et al. 1997; Scheetz et al. 1997), have been identified in dendrites and dendritic spines (Tiedge and Brosius 1996; Steward 1997). While most attention has been paid to postsynaptic protein synthesis in dendrites, there is also evidence for presynaptic targeting of mRNAs, and the synthesis of specific proteins has been identified in axons and axon terminals, particularly in invertebrate neurons (Crispino et al. 1997; Steward 1997; Koenig and Giuditta 1999). Local protein synthesis may be used in several different ways, depending on the developmental stage of the neuron. In immature neurons, protein synthesis in dendritic or axonal growth cones may provide key proteins that are critical for neurite growth or guidance (Steward 1997). The rate of synaptic protein synthesis changes with development and peaks during the period of most active synapse stabilization (Steward and Falk 1986). Local protein synthesis in invertebrate neurons appears to be critical for long-term facilitation and memory storage (Martin et al. 1997; Casadio et al. 1999). In mature vertebrate neurons, local protein synthesis in dendrites may play a key role in processes such as long-term potentiation (LTP) and long-term depression (LTD). Typical neurons in the central nervous system contain as many as 1000 synapses, and evidence exists that segregated groups of synapses within the same neuron can be potentiated independently (Schuman 1997). The discovery of local protein synthesis in dendrites provides a mechanism by which site-specific changes in protein production could be coupled to synapse-specific plasticity.

During the development of many central nervous system pathways, the activation of NMDA glutamate receptors mediates synaptic competition and stabilization. This process has been extensively studied in the amphibian tadpole retinotectal projection, and previous studies had found that certain unidentified proteins were phosphorylated in tadpole tecta following NMDA receptor activation (Scheetz and Constantine-Paton 1996). The most prominent of these proteins has been recently identified as eEF2 (Scheetz et al. 1997). Phosphorylation of eEF2 is increased by patterned visual stimulation of the retina in dark-adapted tadpoles in retino-recipient layers of the tectum that receive visual input. Using the phospho-eEF2 antibody, light microscope analysis indicated that glutamate treatment of tadpole tecta increased the phosphorylation of the protein in all compartments of stimulated cells. In contrast, in studies performed in adult frogs, glutamate-induced phosphorylation of eEF2 was restricted to a small region of the tectum that is the site of binocular innervation from the opposite visual field. Notably, this region (layer 9a) contains no neuronal cell bodies, but only distal dendritic segments. A more detailed analysis using immuno-electron microscopy revealed that the phospho-eEF2 signal was restricted to the postsynaptic aspect of the synapse, in dendritic spines and shafts. Thus, phosphorylation of eEF2 is likely to play an important role in the regulation of local protein synthesis in neuronal dendrites.

11
eEF2 Phosphorylation and Glutamate-Mediated Control of Protein Synthesis at Developing Synapses

Recent evidence also indicates that synaptic activity, or activation of neurotransmitter receptors, can regulate local protein synthesis. In hippocampal slices, NMDA and acetylcholine rapidly increased the synthesis of proteins in dendrites (Feig and Lipton 1993). In synaptoneurosomes, a postsynaptic preparation, synthesis of the fragile X mental retardation protein was increased by activation of metabotropic glutamate receptors (Weiler and Greenough 1993; Weiler et al. 1997), and depolarization increased the synthesis of several unidentified proteins in a Ca^{2+}-dependent manner (Leski and Steward 1996). In hippocampal slices, neurotrophic factors also caused a rapid and long-lasting enhancement of synaptic strength, and this was dependent on immediate protein synthesis (Kang and Schuman 1996). Notably, synaptic activation of CA1 neurons in the hippocampus increased the synthesis of CaM kinase II (Ouyang et al. 1997; Steward and Halpain 1999), a protein known to be involved in the regulation of synaptic plasticity. The observation that eEF2 phosphorylation is spatially restricted to subsynaptic sites in mature neurons raised a number of interesting possibilities. As discussed above, NMDA receptor activation and local protein synthesis have been implicated in forms of synaptic plasticity that involve long-term changes in neuronal structure, function and protein expression (Schuman 1997; Steward 1997). The fact that eEF2 phosphorylation is regulated in dendrites in response to NMDA receptor activation may provide an important Ca^{2+}-dependent link to the regulation of local protein synthesis in neuronal dendrites.

In recent studies it has been observed that brief NMDA treatment of synaptoneurosomes leads to the rapid phosphorylation of eEF2 (Scheetz et al. 2000). As observed using cultured neurons, this NMDA treatment also led to the inhibition of overall protein synthesis that was maximal after 10 min of stimulation, but then returned to basal levels. Using two-dimensional isoelectric focusing/SDS-PAGE, a limited group of proteins were found to be newly synthesized under basal conditions or during the period of maximal inhibition of protein synthesis, but a much larger array of newly synthesized proteins appeared at later time points. These results suggest that brief NMDA receptor activation regulates the translation of mRNA species that are present in synaptoneurosomes. Only a restricted number of mRNA species are targeted to dendrites, and are likely to be present in synaptoneurosomes. One such mRNA is that for the α-subunit of CaM kinase II (Mayford et al. 1996), and the level of CaM kinase IIα protein has been shown to increase rapidly in dendrites following tetanic stimulation of CA1 neurons in hippocampal slices (Ouyang et al. 1997). Using antibodies specific for CaM kinase II, it was found that the level of CaM kinase IIα protein increases within 5 min after treatment with NMDA, the period during which total protein synthesis is decreased. In support of a role for the inhibition of elongation, low concentrations of cycloheximide, sufficient to inhibit total protein synthesis by about 60%, resulted in an increase in synthesis of CaM kinase II of over 250%.

Taken together, these results suggest that eEF2 phosphorylation or cycloheximide treatment results in a decrease in elongation rate that is sufficient to account for both the decrease in total protein synthesis as well as the rapid increase in synaptic CaM kinase IIα synthesis. A similar resistance to the effects of cycloheximide has been observed in other studies of CaM kinase II synthesis in hippocampal neuronal dendrites following synaptic activation (Steward and Halpain 1999). Other studies using synaptoneurosomes have found that glutamate treatment or membrane depolarization results in an increase in the association of CaM kinase IIα mRNA, but not other mRNAs, with polysomes and that CaM kinase II synthesis increases (Bagni et al. 2000).

The mechanism(s) whereby glutamate treatment is coupled to preferential association of CaM kinase IIα mRNA with polysomes is not clear but may involve eEF2 phosphorylation and inhibition of elongation. In other studies of mRNA competition, the efficiency of translation of certain transcripts is paradoxically increased by the inhibition of elongation using cycloheximide (Brendler et al. 1981a, b; Godefroy-Colburn and Thach 1981; Walden et al. 1981). For example, in fibroblasts, doses of cycloheximide that reduce overall protein synthesis by 50% result in the remaining synthesis being accounted for by only a few proteins (Walden and Thach 1986). This may occur as a result of the inhibition of elongation favoring translation of mRNAs that normally exhibit low initiation rates. In addition to eEF2 phosphorylation and inhibition of elongation, other processes are likely to be involved in the regulation of CaM kinase II synthesis in dendrites. The targeting of CaM kinase IIα mRNA to the dendrite is likely to be regulated. Recently, it has been shown in rat visual cortex that cytoplasmic polyadenylation of CaM kinase IIα mRNA is increased by light after dark-rearing of the animals (Wu et al. 1998). This effect is correlated with an increase in synthesis of CaM kinase II protein levels at the synapse, suggesting that polyadenylation increases translational efficiency in neurons in a similar manner to that found in oocytes (Richter 1999).

12 Concluding Remarks Concerning the Physiological Role of eEF2 Phosphorylation

The results discussed in this review support the idea that phosphorylation of eEF2 by EF2 kinase plays an important role in the regulation of protein synthesis at the level of elongation and that this process may be involved in specific aspects of cell function that are modulated by Ca^{2+}. eEF2 phosphorylation is increased transiently in response to a variety of agents that raise intracellular Ca^{2+} levels, and this is associated with a transient inhibition of protein synthesis. eEF2 is also phosphorylated in a cell cycle-dependent manner, and may contribute to the inhibition of protein synthesis during mitosis. Other studies indicate that eEF2 phosphorylation is regulated by one or more components of the rapamycin-sensitive signaling pathway. Thus, eEF2 phosphorylation is high in quiescent cells, or in cells in which the mTOR pathway is inhibited. Phosphorylation of eEF2 is functionally equivalent to decreasing the level of the factor, albeit in a transient manner. Taken together with the results that indicate that the biosynthesis of the protein is a major target of rapamycin-mediated inhibition of mTOR, it is likely that regulation of the functional level of eEF2 by de novo synthesis or phosphorylation plays an important role in cell growth and proliferation.

In addition to a role in the overall control of protein translation, the dynamic regulation of phosphorylation of eEF2, that is coupled to changes in intracellular Ca^{2+}, may play a more specific function. Transient phosphorylation of eEF2 and inhibition of protein synthesis could act as a stalling device that is coupled in some way to the specific synthesis of mRNAs that are transcribed in response to Ca^{2+} signals. This process might therefore serve to synchronize transcription and translation and act to amplify a specific signal. Alternatively, transient inhibition of protein synthesis via eEF2 phosphorylation might lead to the disappearance of short-lived proteins that act as regulators of protein expression, either at a transcriptional or translational level. An interesting possibility is that eEF2 phosphorylation constitutes a cellular equivalent of cycloheximide. Notably, blockage of elongation by submaximal concentrations of cycloheximide, while inhibiting general protein synthesis, has been shown to cause the specific translation of certain mRNAs which exhibit low initiation rates. Support for such a role for eEF2 phosphorylation in the selective translation of specific mRNA species has recently been obtained in studies of protein synthesis at developing synapses in the central nervous system.

Finally, within cells, Ca^{2+} levels are regulated in a precise spatial and temporal fashion. Thus, the cellular pool of eEF2 that is phosphorylated may be highly localized. In this respect, a transient Ca^{2+}-dependent inhibition of elongation could modulate protein synthesis in specific subcellular compartments in a precise temporal fashion. Alternatively, Ca^{2+}-dependent inhibition of elongation may occur at certain times or places in cells when other Ca^{2+}-dependent processes such as secretion or contraction are stimulated by hormones or neurotransmitters. Recent studies in neurons have provided support for the idea that eEF2 phosphorylation plays a role in the regulation of this so-called local protein synthesis. Future studies that make use of recently developed inhibitors of EF2 kinase, or of cell or animal systems in which the EF2 kinase gene has been knocked-out, will hopefully provide additional information about the physiological function of the regulation of elongation through eEF2 phosphorylation.

Acknowledgements. This work was supported by USPHS grant GM50402 (A.C.N.).

References

Aberle H, Bauer A, Stappert J, Kispert A, Kemler R (1997) Beta-catenin is a target for the ubiquitin-proteosome pathway. EMBO J 16:3797–3804

Abraham RT (1998) Mammalian target of rapamycin: immunosuppressive drugs uncover a novel pathway of cytokine receptor signaling. Curr Opin Immunol 10:330–336

Alirezaei M, Nairn AC, Glowinski J, Premont J, Marin P (1999) Zinc inhibits protein synthesis in neurons: potential role of the phosphorylation of the translation initiation factor-2. J Biol Chem 274:32433–32438

Alirezaei M, Marin P, Nairn AC, Glowinski J, Premont J (2001) Hydrogen peroxide inhibits protein synthesis in cortical neurons. J Neurochem 76:1080–1088

Anraku Y, Ohya Y, Iida H (1991) Cell cycle control by calcium calmodulin in *Saccharomyces cerevisiae.* Biochim Biophys Acta 1093:169–177

Avruch J (1998) Insulin signal transduction through protein kinase cascades. Mol Cell Biochem 182:31–48

Bagaglio DM, Cheng EHC, Gorelick FS, Mitsui K, Nairn AC, Hait WN (1993) Phosphorylation of elongation factor 2 in normal malignant rat glial cells. Cancer Res 53:2260–2264

Bagni C, Mannucci L, Dotti CG, Amaldi F (2000) Chemical stimulation of synaptosomes modulates alpha -Ca^{2+}/calmodulin-dependent protein kinase II mRNA association to polysomes. J Neurosci (online) 20:RC76

Ban N, Nissen P, Hansen J, Capel M, Moore PB, Steitz TA (1999) Placement of protein RNA structures into a 5 A-resolution map of the 50 S ribosomal subunit. Nature 400:841–847

Berridge MJ (1994) The biology medicine of calcium signalling. Mol Cell Endocrinol 98:119–124

Berridge MJ (1998) Neuronal calcium signaling. Neuron 21:13–26

Bonneau A-M, Sonenberg N (1987) Involvement of the 24-kDa cap-binding protein in regulation of protein synthesis in mitosis. J Biol Chem 262:11134–11139

Brady MJ, Nairn AC, Wagner JA, Palfrey HC (1990) Nerve growth factor-induced downregulation of calmodulin-dependent protein kinase III in PC12 cells involves cAMP-dependent protein kinase. J Neurochem 54:1034–1039

Braun AP, Schulman H (1995) The multifunctional calcium/calmodulin-dependent protein kinase: from form to function. Annu Rev Physiol 57:417–445

Brendler T, Godefroy-Colburn T, Carlill RD, Thach RE (1981a) The role of mRNA competition in regulating translation. II. Development of a quantitative in vitro assay. J Biol Chem 256:11747–11754

Brendler T, Godefroy-Colburn T, Yu S, Thach RE (1981b) The role of mRNA competition in regulating translation. III. Comparison of in vitro and in vivo results. J Biol Chem 256:11755–11761

Brostrom CO, Chin KV, Wong WL, Cade C, Brostrom MA (1989) Inhibition of translational initiation in eukaryotic cells by calcium ionophore. J Biol Chem 264:1644–1649

Brunn GJ, Hudson CC, Sekulic A, Williams JM, Hosoi H, Houghton PJ, Lawrence JC Jr, Abraham RT (1997) Phosphorylation of the translational repressor PHAS-I by the mammalian target of rapamycin. Science 277:99–101

Burgin KE, Waxham MN, Rickling S, Westgate SA, Mobley SC, Kelly PT (1990) In situ hybridization and histochemistry of Ca^{2+}/calmodulin-dependent protein kinase in developing rat brain. J Neurosci 10:1788–1798

Campbell LE, Wang X, Proud CG (1999) Nutrients differentially regulate multiple translation factors and their control by insulin. Biochem J 344:433–441

Carlberg U, Nilsson A, Nygård O (1990) Functional properties of phosphorylated elongation factor 2. Eur J Biochem 191: 639–645

Carlberg U, Nilsson A, Skog S, Palmquist K, Nygård O (1991) Increased activity of the eEF-2 specific Ca^{2+} calmodulin dependent protein kinase III during the S-phase in Ehrlich ascites cells. Biochem Biophys Res Commun 180:1372–1376

Casadio A, Martin KC, Giustetto M, Zhu H, Chen M, Bartsch D, Bailey CH, Kandel ER (1999) A transient neuron-wide form of CREB-mediated long-term facilitation can be stabilized at specific synapses by local protein synthesis. Cell 99:221–237

Celis JE, Madsen P, Ryazanov AG (1990) Increased phosphorylation of elongation factor 2 during mitosis in transformed human amnion cells correlates with a decreased rate of protein synthesis. Proc Natl Acad Sci USA 87:4231–4235

Chen J, Peterson RT, Schreiber SL (1998) α4 associates with protein phosphatases 2A, 4 and 6. Biochem Biophys Res Commun 247:827–832

Chen ZJ, Parent L, Maniatis T (1996) Site-specific phosphorylation of IkappaB by a novel ubiquitination-dependent protein kinase activity. Cell 84:853–862

Cho SI, Koketsu M, Ishihara H, Matsushita M, Nairn AC, Fukazawa H, Uehara Y (2000) Novel compounds and 1,3-selenazine derivatives as specific inhibitors of eukaryotic elongation factor-2 kinase. Biochim Biophys Acta 1475:207–215

Chou MM, Hou W, Johnson J, Graham LK, Lee MH, Chen CS, Newton AC, Schaffhausen BS, Toker A (1998) Regulation of protein kinase C zeta by PI 3-kinase PDK-1. Curr Biol 24:1069–1077

Chung HY, Nairn AC, Murata K, Brautigan DL (1999) Mutation of Tyr307 and Leu309 in the protein phosphatase 2A catalytic subunit favors association with the α4 subunit which promotes dephosphorylation of elongation factor-2. Biochemistry 38:10371–10376

Clemens M (1996) Protein kinases that phosphorylate eIf2 and eIF2B and their role in eukaryotic cell translational control. In: Hershey JWB, Mathews MB, Sonenberg N (eds) Translational control. Cold Spring Harbor Laboratory Press, Cold Spring Harbor, pp 139–172

Côté GP, Luo X, Murphy MB, Egelhoff TT (1997) Mapping of the novel protein kinase catalytic domain of *Dictyostelium* myosin II heavy chain kinase A. J Biol Chem 272:6846–6849

Crispino M, Kaplan BB, Martin R, Alvarez J, Chun JT, Benech JC, Giuditta A (1997) Active polysomes are present in the large presynaptic endings of the synaptosomal fraction from squid brain. J Neurosci 17:7694–7702

DeGarcia DJ, Neumar RW, White BC, Krause GS (1996) Global brain ischemia reperfusion: modifications in eukaryotic initiation factors associated with inhibition of translation initiation. J Neurochem 67:2005–2012

Di Como CJ, Arndt KT (1996) Nutrients via the Tor proteins stimulate the association of Tap42 with type 2 A phosphatases. Genes Dev 10:1904–1916

Diggle TA, Redpath NT, Heesom KJ, Denton RM (1998) Regulation of protein-synthesis elongation-factor-2 kinase by cAMP in adipocytes. Biochem J 336:525–529

Diggle TA, Seehra CK, Hase S, Redpath NT (1999) Analysis of the domain structure of elongation factor-2 kinase by mutagenesis: FEBS Lett 457:189–192

Dumont-Miscopein A, Lavergne J-P, Guillot D, Sontag B, Reboud J-P (1994) Interaction of phosphorylated elongation factor EF-2 with nucleotides and ribosomes. FEBS Lett 356:283–286

Everett AD, Stoops TD, Cho H, Nairn AC, Brautigan D (2001) Angiotensin II regulation of protein synthesis in cardiac myocytes involves elongation factor 2. Amer J Physiol (In Press)

Feig S, Lipton P (1993) Pairing the cholinergic agonist carbachol with patterned Schaffer collateral stimulation initiates protein synthesis in hippocampal CA1 pyramidal cell dendrites via a muscarinic NMDA-dependent mechanism. J Neurosci 13:1010–1021

Feng Y, Gutekunst CA, Eberhart DE, Yi H, Warren ST, Hersch SM (1997) Fragile X mental retardation protein: nucleocytoplasmic shuttling association with somatodendritic ribosomes. J Neurosci 17:1539–1547

Fox HL, Kimball SR, Jefferson LS, Lynch CJ (1998) Amino acid effects on translational repressor 4E-BP1 are mediated primarily by L-leucine in isolated adipocytes. Am J Physiol Cell Physiol 43:C206–213

Frank J, Agrawal RJ (2000) A ratchet-like inter-subunit reorganization of the ribosome during translocation. Nature 406:318–322

Frederickson RM, Mushynski WE, Sonenberg N (1992) Phosphorylation of translation initiation factor eIF-4E is induced in a ras-dependent manner during nerve growth factor-mediated PC12 cell differentiation. Mol Cell Biol 12:1239–1247

Furukawa K, Estus S, Fu W, Mark RJ, Mattson MP (1997) Neuroprotective action of cycloheximide involves induction of bcl-2 antioxidant pathways. J Cell Biol 136:1137–1149

Garner CC, Tucker RP, Matus A (1988) Selective localization of messenger RNA for cytoskeletal protein MAP2 in dendrites. Nature 336:674–677

Gill DM, Dinius LL (1973) The EF-2 content of mammalian cells. J Biol Chem 248:654–658

Gingras AC, Gygi SP, Raught B, Polakiewicz RD, Abraham RT, Hoekstra MF, Aebersold R, Sonenberg N (1999) Regulation of 4E-BP1 phosphorylation: a novel two-step mechanism. Genes Dev 13:1422–1437

Gingras AC, Kennedy SG, O'Leary MA, Sonenberg N, Hay N (1998) 4E-BP1 a repressor of mRNA translation is phosphorylated and inactivated by the Akt(PKB) signaling pathway. Genes Dev 12:502–513

Godefroy-Colburn T, Thach RE (1981) The role of mRNA competition in regulating translation. IV Kinetic model. J Biol Chem 256:11762–11773

Goldberg J, Nairn AC, Kuriyan J (1996) Structural basis for the autoinhibition of calcium calmodulin-dependent protein kinase I. Cell 84:875–887

Green R (2000) Ribosomal translocation: EF-G turns the crank. Curr Biol 10:R369–373

Greenberg ME, Belasco JG (1993) Control of decay of labile protooncogene cytokine mRNAs. In: Belasco JG, Brawerman G (eds) Control of mRNA stability. Academic Press, New York, pp 199–218

Greenberg ME, Hermanowski AL, Ziff EA (1986) Effects of protein synthesis inhibitors on growth factor activation of c-fos, c-myc and actin gene transcription. Mol Cell Biol 6:1050–1057

Groigno L, Whitaker M (1998) An anaphase calcium signal controls chromosome disjunction in early sea urchin embryos. Cell 92:193–204

Gwag BJ, Canzoniero LMT, Sensi SL, Demaro JA, Koh JY, Goldberg MP, Jacquin M, Choi DW (1999) Calcium ionophores can induce either apoptosis or necrosis in cultured cortical neurons. Neuroscience 90:1339–1348

Hara K, Yonezawa K, Weng QP, Kozlowski MT, Belham C, Avruch J (1998) Amino acid sufficiency and mTOR regulate p70 S6 kinase and eIF- 4E BP1 through a common effector mechanism. J Biol Chem 273:14484–14494

Harding HP, Zhang Y, Bertolotti A, Zeng H, Ron D (2000) Perk is essential for translational regulation and cell survival during the unfolded protein response. Mol Cell 5:897–904

Harteneck C, Plant TD, Schultz G (2000) From worm to man: three subfamilies of TRP channels. Trends Neurosci 23:159–166

Hashemolhosseini S, Nagamine Y, Morley SJ, Desrivieres S, Mercep L, Ferrari S (1998) Rapamycin inhibition of the G1 to S transition is mediated by effects on cyclin D1 mRNA protein stability. J Biol Chem 273:14424–14429

Hazelrigg T (1998) The destinies and destinations of RNAs. Cell 95:451–460

Hershko A (1997) Roles of ubiquitin-mediated proteolysis in cell cycle control. Curr Opin Cell Biol 9:788–799

Hershko A, Ciechanover A (1998) The ubiquitin system. Annu Rev Biochem 67:425–479

Hincke MT, Nairn AC (1992) Phosphorylation of elongation factor 2 during Ca^{2+}-mediated secretion from rat parotid acini. Biochem J 282:877–882

Hon WC, McKay GA, Thompson PR, Sweet RM, Yang DSC, Wright GD, Berghuis AM (1997) Structure of an enzyme required for aminoglycoside antibiotic resistance reveals homology to eukaryotic protein kinases. Cell 89:887–895

Hovland R, Eikhom TS, Proud CG, Cressey LI, Lanotte M, Doskeland SO, Houge G (1999) cAMP inhibits translation by inducing Ca^{2+}/calmodulin-independent elongation factor 2 kinase activity in IPC-81 cells. FEBS Lett 444:97–101

Hu BR, Wieloch T (1993) Stress-induced inhibition of protein synthesis initiation: modulation of initiation factor 2 guanine nucleotide exchange factor activities following transient cerebral ischemia in the rat. J Neurosci 13:1830–1838

Jaye M, Schlessinger J, Dionne CA (1992) Fibroblast growth factor receptor tyrosine kinases: molecular analysis signal transduction. Biochim Biophys Acta 1135:185–199

Jefferies HBJ, Thomas G (1996) Ribosomal protein S6 phosphorylation signal transduction. In: Hershey JWB, Mathews MB, Sonenberg N (eds) Translational control. Cold Spring Harbor Laboratory Press, Cold Spring Harbor, pp 389–410

Jefferies HBJ, Reinhard C, Kozma SC, Thomas G (1994) Rapamycin selectively represses translation of the "polypyrimidine tract" mRNA family. Proc Natl Acad Sci USA 91:4441–4445

Jefferies HB, Fumagalli S, Dennis PB, Reinhard C, Pearson RB, Thomas G (1997) Rapamycin suppresses 5′TOP mRNA translation through inhibition of $p70^{S6k}$. EMBO J 16:3693–3704

Jiang Y, Broach JR (1999) Tor proteins and protein phosphatase 2 A reciprocally regulate Tap42 in controlling cell growth in yeast. EMBO J 18:2782–2792

Kang HJ, Schuman EM (1996) A requirement for local protein synthesis in neurotrophin-induced hippocampal synaptic plasticity. Science 273:1402–1406

Kanki JP, Newport JW (1991) The cell cycle dependence of protein synthesis during *Xenopus laevis* development. Dev Biol 146:198–213

Kleiman R, Banker G, Steward O (1990) Differential subcellular localization of particular mRNA in hippocampal neurons in culture. Neuron 5:821–830

Koenig E, Giuditta A (1999) Protein synthesizing machinery in the axon compartment. Neuroscience 89:5–15

Kohno K, Uchida T (1987) Highly frequent single amino acid substitution in mammalian elongation factor 2 (EF-2) results in expression of resistance to EF-2-ADP-ribosylating toxins. J Biol Chem 262:12298–12305

Kohno K, Uchida T, Ohkubo H, Nakanishi S, Nakanishi T, Fukui T, Ohtsuka E, Ikehara M, Okada Y (1986) Amino acid sequence of mammalian elongation factor 2 deduced from the cDNA sequence: homology with GTP-binding proteins. Proc Natl Acad Sci USA 83:4978–4982

Kolman MF, Egelhoff TT (1997) Dictyostelium myosin heavy chain kinase A subdomains – coiled-coil WD repeat roles in oligomerization substrate targeting. J Biol Chem 272: 16904–16910

Laitusis AL, Brostrom CO, Ryazanov AG, Brostrom MA (1998) An examination of the role of increased cytosolic free Ca^{2+} concentrations in the inhibition of mRNA translation. Arch Biochem Biophys 354:270–280

Lazaris-Karatzas A, Montine KS, Sonenberg N (1990) Malignant transformation by a eukaryotic initiation factor subunit that binds to mRNA 5′ cap. Nature 345:544–547

Le Good JA, Ziegler WH, Parekh DB, Alessi DR, Cohen P, Parker PJ (1998) Protein kinase C isotypes controlled by phosphoinositide 3-kinase through the protein kinase PDK1. Science 281:2042–2045

Leski ML, Steward O (1996) Protein synthesis within dendrites: ionic neurotransmitter modulation of synthesis of particular polypeptides characterized by gel electrophoresis. Neurochem Res 21:681–690

Levenson RM, Nairn AC, Blackshear PJ (1989) Insulin rapidly induces the biosynthesis of elongation factor 2. J Biol Chem 264:11904–11911

Lu KP, Means AR (1993) Regulation of the cell cycle by calcium calmodulin. Endocrinol Rev 14:40–58

Lu KP, Osmani SA, Osmani AH, Means AR (1993) Essential roles for calcium calmodulin in G2/M progression in *Aspergillus nidulans.* J Cell Biol 121:621–630

Mackie KP, Nairn AC, Hampel G, Lam G, Jaffe EA (1989) Thrombin and histamine stimulate the phosphorylation of elongation factor 2 in endothelial cells. J Biol Chem 264:1748–1753

Marin P, Nastiuk KL, Daniel N, Girault JA, Czernik AJ, Glowinski J, Nairn AC, Prémont J (1997) Glutamate-dependent phosphorylation of elongation factor-2 inhibition of protein synthesis in neurons. J Neurosci 17:3445–3454

Martin KC, Casadio A, Zhu H, Rose JC, Chen M, Bailey CH, Kandel ER (1997) Synapse-specific long-form facilitation of Aplysia sensory to motor synapses: a function for local protein synthesis in memory storage. Cell 91:927–938

Mayford M, Baranes D, Podsypanina K, Kandel ER (1996) The 3′-untranslated region of CaMKIIα is a *cis*-acting signal for the localization and translation of mRNA in dendrites. Proc Natl Acad Sci USA 93:13250–13255

Means AR, Rasmussen CD (1988) Calcium calmodulin cell proliferation. Cell Calcium 9:313–319

Mendez R, Kollmorgen G, White MF, Rhoads RE (1997) Requirement of protein kinase Cxi for stimulation of protein synthesis by insulin. Mol Cell Biol 17:5184–5192

Merrick WC, Hershey JWB (1996) The pathway mechanism of eukaryotic protein synthesis. In: Hershey JWB, Mathews MB, Sonenberg N (eds) Translational control. Cold Spring Harbor Laboratory Press, Cold Spring Harbor, pp 31–70

Mitsui K, Brady M, Palfrey HC, Nairn AC (1993) Purification and characterization of calmodulin-dependent protein kinase III from rabbit reticulocytes and rat pancreas. J Biol Chem 268:13422–13433

Mitsui K, Lan M, Paredes M, Matsushita M, Jaffe EA, Palfrey HC, Nairn AC (2001) Ca^{2+}-dependent regulation of protein synthesis by elongation factor 2 phosphorylation in intact cells. J Biol Chem (In revision)

Miyashiro K, Dichter M, Eberwine J (1994) On the nature and differential distribution of mRNAs in hippocampal neurites: implications for neuronal functioning. Proc Natl Acad Sci USA 91:10800–10804

Muise-Helmericks RC, Grimes HL, Bellacosa A, Malstrom SE, Tsichlis PN, Rosen N (1998) Cyclin D expression is controlled post-transcriptionally via a phosphatidylinositol 3-Kinase/ Akt-dependent pathway. J Biol Chem 273:29864–29872

Murata K, Wu J, Brautigan DL (1997) B cell receptor-associated protein α4 displays rapamycin-sensitive binding directly to the catalytic subunit of protein phosphatase 2 A. Proc Natl Acad Sci USA 94:10624–10629

Nairn AC, Palfrey HC (1987) Identification of the major Mr 100 000 substrate for calmodulin-dependent protein kinase III in mammalian cells as elongation factor-2 J. Biol Chem 262:17299–17303

Nairn AC, Palfrey HC (1996) Regulation of protein synthesis by calcium In: Hershey JWB, Mathews MB, Sonenberg N (eds) Translational control. Cold Spring Harbor Laboratory Press, Cold Spring Harbor, pp 295–318

Nairn AC, Bhagat B, Palfrey HC (1985) Identification of calmodulin-dependent protein kinase III and its major 100 000-molecular-weight substrate in mammalian tissues. Proc Natl Acad Sci USA 82: 7939–7943

Nairn AC, Nichols RA, Brady MJ, Palfrey HC (1987) Nerve growth factor treatment or cyclic AMP elevation reduces calcium-calmodulin-dependent protein kinase III activity in PC12 cells. J Biol Chem 262: 14265–14272

Nilsson L, Nygård O (1991) Altered sensitivity of eukaryotic elongation factor 2 for trypsin after phosphorylation and ribosomal binding. J Biol Chem 266:10578–10582

Nilsson A, Nygård O (1995) Phosphorylation of eukaryotic elongation factor 2 in differentiating and proliferating HL-60 cells. Biochim Biophys Acta 1268:263–268

Nilsson A, Carlberg U, Nygård O (1991) Kinetic characterisation of the enzymatic activity of the eEF-2-specific Ca^{2+}-calmodulin-dependent protein kinase III purified from rabbit reticulocytes. Eur J Biochem 195:377–383

Nygård O, Nilsson L (1984) Quantification of the different ribosomal phases during the translational elongation cycle in rabbit reticulocyte lysates. Eur J Biochem 145:345–350

Nygård O, Nilsson A, Carlberg U, Nilsson L, Amons R (1991) Phosphorylation regulates the activity of the eEF-2-specific Ca^{2+}-calmodulin-dependent protein kinase III. J Biol Chem 266:16425–16430

Orrego F, Lipmann F (1967) Protein synthesis in brain slices. J Biol Chem 242:665–671

Ouyang Y, Kantor D, Harris KM, Schuman EM, Kennedy MB (1997) Visualization of the distribution of autophosphorylated calcium/calmodulin-dependent protein kinase II after tetanic stimulation in the CA1 area of the hippocampus. J Neurosci 17:5416–5427

Ovchinnikov LP, Motuz LP, Natapov PG, Averbuch LJ, Wettenhall REH, Szyszka R, Kramer G, Hardesty B (1990) Three phosphorylation sites in elongation factor 2. FEBS Lett 275:209–212

Palfrey HC (1983) Presence in many mammalian tissues of an identical major cytosolic substrate (Mr 100 000) for calmodulin-dependent protein kinase. FEBS Lett 157:183–190

Palfrey HC, Nairn AC (1995) Calcium-dependent regulation of protein synthesis. Adv Second Messenger Phosphoprotein Res 30:191–223

Palfrey HC, Nairn AC, Muldoon LL, Villereal ML (1987) Rapid activation of calmodulin-dependent protein kinase III in mitogen-stimulated human fibroblasts: correlation with intracellular Ca^{2+} transients. J Biol Chem 262: 9785–9792

Peters KG, Marie J, Wilson E, Ives HE, Escobedo J, Del Rosario M, Mirda D, Williams LT (1992) Point mutation of an FGF receptor abolishes phosphatidylinositol turnover and Ca^{2+} flux but not mitogenesis. Nature 358:678–681

Pike BR, Zhao X, Newcomb JK, Wang KK, Posmantur RM, Hayes RL (1998) Temporal relationships between de novo protein synthesis and calpain caspase 3-like protease activation and DNA fragmentation during apoptosis in septo-hippocampal cultures. J Neurosci Res 52:505–520

Poenie M, Alderton J, Tsien RY, Steinhardt RA (1985) Changes of free calcium levels with stages of the cell division cycle. Nature 315:147–149

Poenie M, Alderton J, Steinhardt R, Tsien R (1986) Calcium rises abruptly and briefly throughout the cell at the onset of anaphase. Science 233:886–889

Polekhina G, Thirup S, Kjeldgaard M, Nissen P, Lippmann C, Nyborg J (1996) Helix unwinding in the effector region of elongation factor EF-Tu-GDP. Structure 4:1141–1151

Price NT, Redpath NT, Severinov KV, Campbell DG, Russell JM, Proud CG (1991) Identification of the phosphorylation sites in elongation factor- 2 from rabbit reticulocytes. FEBS Lett 282:253–258

Prostko CR, Brostrom MA, Brostrom CO (1993) Reversible phosphorylation of eIF-2alpha in response to ER signalling. Mol Cell Biochem 127/8:255–265

Proud CG (1992) Protein phosphorylation in translational control. Curr Topics Cell Regul 32:243–369

Racca C, Gardiol A, Triller A (1997) Dendritic postsynaptic localizations of glycine receptor alpha subunit mRNAs. J Neurosci 17:1691–1700

Raley-Susman KM, Lipton P (1990) In vitro ischemia protein synthesis in rat hippocampal slices: the role of calcium NMDA. Brain Res 515:27–38

Rattan SI (1991) Protein synthesis and the components of the protein synthetic machinery during cellular aging. Mutation Res 256:115–125

Redpath NT, Proud CG (1989) The tumour promoter okadaic acid inhibits reticulocyte-lysate protein synthesis by increasing the net phosphorylation of elongation factor 2. Biochem J 262:69–75

Redpath NT, Proud CG (1990) Activity of protein phosphatases against initiation factor-2 and elongation factor-2. Biochem J 272:175–180
Redpath NT, Proud CG (1993a) Purification and phosphorylation of elongation factor-2 kinase from rabbit reticulocytes. Eur J Biochem 212:511–520
Redpath NT, Proud CG (1993b) Cyclic AMP-dependent protein kinase phosphorylates rabbit reticulocyte elongation factor-2 kinase and induces calcium-independent activity. Biochem J 293:31–34
Redpath NT, Proud CG (1994) Molecular mechanisms in the control of translation by hormones and growth factors. Biochim Biophys Acta Mol Cell Res 1220:147–162
Redpath NT, Price NT, Severinov KV, Proud CG (1993) Regulation of elongation factor-2 by multisite phosphorylation. Eur J Biochem 213:689–699
Redpath NT, Price NT, Proud CG (1996a) Cloning expression of cDNA encoding protein synthesis elongation factor-2 kinase. J Biol Chem 271:17547–17554
Redpath NT, Foulstone EJ, Proud CG (1996b) Regulation of translation elongation factor-2 by insulin via a rapamycin-sensitive signalling pathway. EMBO J 15:2291–2297
Rhoads RE (1993) Regulation of eukaryotic protein synthesis by initiation factors. J Biol Chem 268:3017–3020
Richter JD (1999) Cytoplasmic polyadenylation in development and beyond. Microbiol Mol Biol Rev 63:446–456
Rinker-Schaeffer CW, Austin V, Zimmer S, Rhoads RE (1992) ras transformation of cloned rat embryo fibroblasts results in increased rates of protein synthesis phosphorylation of eukaryotic initiation factor 4 E. J Biol Chem 267:10659–10664
Rodnina MV, Savelsbergh A, Katunin VI, Wintermeyer W (1997) Hydrolysis of GTP by elongation factor G drives tRNA movement on the ribosome. Nature 385:37–41
Ross J (1997) A hypothesis to explain why translation inhibitors stabilize mRNAs in mammalian cells: mRNA stability mitosis. BioEssays 19:527–529
Ryazanov AG (1987) Ca^{2+}/calmodulin-dependent phosphorylation of elongation factor 2. FEBS Lett 214:331–334
Ryazanov AG, Davydova EK (1989) Mechanism of elongation factor 2 (EF-2) inactivation upon phosphorylation: phosphorylated EF-2 is unable to catalyze translocation. FEBS Lett 251:187–190
Ryazanov AG, Shestakova EA, Natapov PG (1988) Phosphorylation of elongation factor 2 by EF-2 kinase affects rate of translation. Nature 334:170–173
Ryazanov AG, Ward MD, Mendola CE, Pavur KS, Dorovkov MV, Wiedmann M, Erdjument-Bromage H, Tempst P, Parmer TG, Prostko CR, Germino FJ, Hait WN (1997) Identification of a new class of protein kinases represented by eukaryotic elongation factor-2 kinase. Proc Natl Acad Sci USA 94:4884–4889
Ryazanov AG, Pavur KS, Dorovkov MV (1999) Alpha-kinases: a new class of protein kinases with a novel catalytic domain. Curr Biol 9:R43–45
Scheetz AJ, Constantine-Paton M (1996) NMDA receptor activation-responsive phosphoproteins in the developing optic tectum. J Neurosci 16:1460–1469
Scheetz AJ, Nairn AC, Constantine-Paton M (1997) N-methyl-D-aspartate receptor activation and visual activity induce elongation factor-2 phosphorylation in amphibian tecta: a role for *N*-methyl-D-aspartate receptors in controlling protein synthesis. Proc Natl Acad Sci USA 94:14770–14775
Scheetz AJ, Nairn AC, Constantine-Paton M (2000) NMDA receptor-mediated control of protein synthesis at developing synapses. Nat Neurosci 3:211–216
Scheidereit C (1998) Docking IkappaB kinases. Nature 395:225–226
Schlossmann J, Ammendola A, Ashman K, Zong XG, Huber A, Neubauer G, Wang GX, Allescher HD, Korth M, Wilm M, Hofmann F, Ruth P (2000) Regulation of intracellular calcium by a signalling complex of IRAG IP_3 receptor cGMP kinase Iβ. Nature 404:197–201
Schuman EM (1997) Synapse specificity and long-term information storage. Neuron 18:339–342
Severinov KV, Melnikova IG, Ryazanov AG (1990) Downregulation of the translational EF-2alpha kinase in *Xenopus laevis* oocytes at the final stages of oogenesis. New Biol 2:887–893

Shi Y, Vattem KM, Sood R, An J, Liang J, Stramm L, Wek RC (1998) Identification and characterization of pancreatic eukaryotic initiation factor 2 alpha-subunit kinase PEK involved in translational control. Mol Cell Biol 18:7499–7509

Sonenberg N (1996) mRNA 5′ cap-binding protein eIF4 E control of cell growth. In: Hershey JWB, Mathews MB, Sonenberg N (eds) Translational control. Cold Spring Harbor Laboratory Press, Cold Spring Harbor, pp 245–270

Sonenberg N, Gingras AC (1998) The mRNA 5′ cap-binding protein eIF4 E control of cell growth. Curr Opin Cell Biol 10:268–275

Sprinzl M (1994) Elongation factor Tu: a regulatory GTPase with an integrated effector. Trends Biochem Sci 19:245–250

Srivastava SP, Davies MV, Kaufman RJ (1995) Calcium depletion from the endoplasmic reticulum activates the double-stranded RNA-dependent protein kinase (PKR) to inhibit protein synthesis. J Biol Chem 270:16619–16624

Stark H, Rodnina MV, Wieden HJ, van Heel M, Wintermeyer W (2000) Large-scale movement of elongation factor G and extensive conformational change of the ribosome during translocation. Cell 100:301–309

Steward O (1997) mRNA localization in neurons: a multipurpose mechanism. Neuron 18:9–12

Steward O, Falk PM (1986) Protein-synthetic machinery at postsynaptic sites during synaptogenesis. J Neurosci 6:412–423

Steward O, Halpain S (1999) Lamina-specific synaptic activation causes domain-specific alterations in dendritic immunostaining for MAP2 CAM kinase II. J Neurosci 19:7834–7845

Taylor DR, Shi ST, Romano PR, Barber GN, Lai MM (1999) Inhibition of the interferon-inducible protein linase PKR by HCV E2 protein. Science 285:107–110

Terada N, Patel HR, Takase K, Kohno K, Nairn AC, Gelfand EW (1994) Rapamycin selectively inhibits translation of mRNAs encoding elongation factors ribosomal proteins. Proc Natl Acad Sci USA 91:11477–11481

Terada N, Takase K, Papst P, Nairn AC, Gelfand EW (1995) Rapamycin inhibits ribosomal protein synthesis and induces G1 prolongation in mitogen-activated T lymphocytes. J Immunol 155:3418–3426

Thilmann R, Xie Y, Kleihues P, Kiessling M (1986) Persistent inhibition of protein synthesis precedes delayed neuronal death in postischemic gerbil hippocampus. Acta Neuropathol 71:88–93

Tiedge H, Brosius J (1996) Translational machinery in dendrites of hippocampal neurons in culture. J Neurosci 16:7171–7181

Toescu EC (1998) Apoptosis cell death in neuronal cells: where does Ca^{2+} fit in? Cell Calcium 24:387–403

Too CKL (1997) Differential expression of elongation factor-2 α4 phosphoprotein Cdc5-like protein in prolactin-dependent/independent rat lymphoid cells. Mol Cell Endocrinol 131:221–232

Török K, Wilding M, Groigno L, Patel R, Whitaker M (1998) Imaging the spatial dynamics of calmodulin activation during mitosis. Curr Biol 8:692–699

Vary TC, Nairn A, Lynch CJ (1994) Role of elongation factor 2 in regulating peptide-chain elongation in the heart. Am J Physiol Endocrinol Metab 266:E628–E634

Walden WE, Godefroy-Colburn T, Thach RE (1981) The role of mRNA competition in regulating translation. I. Demonstration of competition in vivo. J Biol Chem 256:11739–11746

Walden WE, Thach RE (1986) Translational control of gene expression in a normal fibroblast: characterization of a subclass of mRNAs with unusual kinetic properties. Biochemistry 25:2033–2041

Wang L, Wang X, Proud CG (2000) Activation of mRNA translation in rat cardiac myocytes by insulin involves multiple rapamycin-sensitive steps. Am J Physiol Heart Circ Physiol 278:H1056–1068

Wang X, Campbell LE, Miller CM, Proud CG (1998) Amino acid availability regulates p70 S6 kinase multiple translation factors. Biochem J 334:261–267

Weiler IJ, Greenough WT (1993) Metabotropic glutamate receptors trigger postsynaptic protein synthesis. Proc Natl Acad Sci USA 90:7168–7171

Weiler IJ, Irwin SA, Klintsova AY, Spencer CM, Brazelton AD, Miyashiro K, Comery TA, Patel B, Eberwine J, Greenough WT (1997) Fragile X mental retardation protein is translated near synapses in response to neurotransmitter activation. Proc Natl Acad Sci USA 94:5395–5400

West MJ, Stoneley M, Willis AE (1998) Translational induction of the c-myc oncogene via activation of the FRAP/TOR signalling pathway. Oncogene 17:769–780

Whitaker M, Patel R (1990) Calcium cell cycle control. Development 108:525–542

Wilson KS, Noller HF (1998) Mapping the position of translational elongation factor EF-G in the ribosome by directed hydroxyl radical probing. Cell 92:131–139

Wong WL, Brostrom MA, Brostrom CO (1991) Effects of Ca^{2+} ionophore A23187 on protein synthesis in intact rabbit reticulocytes. Int J Biochem 23:605–608

Wong WL, Brostrom MA, Kuznetsov G, Gmitter-Yellen D, Brostrom CO (1993) Inhibition of protein synthesis and early protein processing by thapsigargin in cultured cells. Biochem J 289:71–79

Wu L, Wells D, Tay J, Mendis D, Abbott MA, Barnitt A, Quinlan E, Heynen A, Fallon JR, Richter JD (1998) CPEB-mediated cytoplasmic polyadenylation and the regulation of experience-dependent translation of alpha-CaMKII mRNA at synapses. Neuron 21:1129–1139

Xu G, Marshall CA, Lin T–A, Kwon G, Munivenkatappa RB, Hill JR, Lawrence JC, McDaniel ML (1998) Branched-chain amino acids are essential in the regulation of PHAS-I p70 S6 kinase by pancreatic beta-cells. A possible role in protein translation mitogenic signaling. J Biol Chem 273:4485–4491

Yaron A, Gonen H, Alkalay I, Hatzubai A, Jung S, Beyth S, Mercurio F, Manning AM, Ciechanover A, Ben-Neriah Y (1997) Inhibition of NF-kappa-B cellular function via specific targeting of the I-kappa-B-ubiquitin ligase. EMBO J 16:6486–6494

Yaron A, Hatzubai A, Davis M, Lavon I, Amit S, Manning AM, Andersen JS, Mann M, Mercurio F, Ben-Neriah Y (1999) Identification of the receptor component of the IkBalpha-ubiquitin ligase. Nature 396:590–594

Phosphorylation of Mammalian eIF4E by Mnk1 and Mnk2: Tantalizing Prospects for a Role in Translation

Malathy Mahalingam and Jonathan A. Cooper[1]

1 Introduction

One of the first translation initiation factors shown to be phosphorylated in mitogen-stimulated mammalian cells was eukaryotic translation initiation factor 4E (eIF4E), the mRNA cap-binding protein (Duncan et al. 1987). eIF4E is also phosphorylated in many malignantly transformed cells (Lazaris Karatzas et al. 1990; Frederickson et al. 1991; Rinker Schaeffer et al. 1992; Rosenwald et al. 1993; Graff et al. 1995), and eIF4E overexpression can lead to transformation (De Benedetti and Rhoads1990; Lazaris Karatzas et al. 1990). In many cell types, eIF4E seems to be limiting for translation initiation (De Benedetti and Rhoads 1990; Mader and Sonenberg 1995). These observations led to the hypothesis that eIF4E is regulated by phosphorylation, that phosphorylation increases translation efficiency, and that increased translation of certain mRNAs can lead to growth and, ultimately, contributes to malignant transformation.

More recently, attention has shifted to the eIF4E binding proteins (4EBPs), which are also phosphorylated in mitogen-stimulated cells and serve to sequester eIF4E in inactive complexes in quiescent cells (Duncan et al. 1987; Haghighat et al. 1995). Studies on the 4EBPs have underscored the importance of eIF4E availability for regulating translation initiation rate. However, the role of phosphorylation of eIF4E itself has been a neglected area, in part because of initial confusion over the exact phosphorylation site and which protein kinases perform the phosphorylation.

In this review, we will describe the identification and characterization of the Mnk serine/threonine protein kinases, which can phosphorylate eIF4E at the physiological site, serine 209, in vitro and appear to be instrumental in phosphorylating eIF4E in cells in response to mitogens and environmental stress. We will also discuss the potential role of phosphorylation in regulating eIF4E function.

[1] Fred Hutchinson Cancer Research Center, P.O. Box 19024, 1100 Fairview Avenue North, A2-025 Seattle, Washington 98109-1024, USA

Progress in Molecular and Subcellular Biology, Vol. 27
Robert E. Rhoads (Ed.)

2 Mnks: Discovery and Kinase Activation

Mnk1 (MAP kinase interacting kinase) and Mnk2 were identified during searches for proteins associated with the mitogen-activated protein (MAP) kinase known as ERK2. ERK2 (and the closely-related ERK1) are activated in response to mitogens by a protein kinase cascade (Ahn et al. 1992; Robbins et al. 1992; Nishida and Gotoh 1993). Stresses such as hypoxia, ionizing radiation, and hypertonic solutions activate similar kinase cascades that lead to the activation of other MAP kinases, which fall into two main families: the JNKs and the p38s, also known as SAPK1s and SAPK2s (Cano and Mahadevan 1995; Ichijo 1999). The ERKs, JNKs and p38s have many identified substrates important for cell regulation, including cytoplasmic factors and nuclear transcription factors (Cano and Mahadevan 1995). Indeed, one interesting aspect of ERK function is that the protein is cytoplasmic when inactive, but after activation, a population of ERK translocates to the nucleus (Chen et al. 1992). Nonetheless, cytoplasmic proteins, including cPLA2 and carbamoyl phosphate synthetase, are phosphorylated by ERK during the mitogenic response (Xu et al. 1994; Hirabayashi et al. 1998; Graves et al. 2000).

Mnk1 and Mnk2 were discovered in screens for mammalian proteins that bind to, or are phosphorylated by ERK2. Mnk2 was cloned in a yeast two-hybrid screen which utilized ERK2 as bait (Waskiewicz et al. 1997) and Mnk1 was cloned by homology to Mnk2. Independently, a HeLa expression library screen for substrates of ERK2 identified Mnk1 (Fukunaga and Hunter 1997). Upon subsequent characterization, Mnk1 was shown to bind to, and be a substrate for, both the mitogenically activated ERK, as well as the stress-activated p38 MAP kinases. However, it does not interact detectably with JNK1. The transcript for Mnk1 is widely expressed in all tissue samples tested, the levels being higher in skeletal muscle and liver, and low in brain (Waskiewicz et al. 1997). Mnk2 has 72% amino acid identity with Mnk1 and binds to ERK2. Unlike Mnk1, Mnk2 does not interact with p38 MAP kinase and is only activated by ERK. However, expression levels of Mnk2 parallel those of Mnk1. This suggests a biochemically divergent role for Mnk2 from Mnk1, whereby Mnk1 may serve to integrate signals received from both ERK and p38 MAP kinase and Mnk2 may be more specific for incoming signals from the ERK MAP kinase cascade (Fig. 1).

The original two-hybrid screen showed that the C-terminal regions of Mnk1 and Mnk2 were sufficient to bind with high affinity to ERK2 (Waskiewicz et al. 1997). Binding did not require the kinase activity of ERK2. The binding site on ERK2 has subsequently been mapped, as has the binding site on Mnk1 (Tanoue et al. 2000). Two aspartates (321 and 324 in *Xenopus*) on ERK2 are part of a binding domain (CD domain, for "common docking") that binds to basic regions in a variety of ERK substrates. The CD region is found in ERKs, JNKs and p38s, but the exact sequence varies, perhaps explaining why Mnk1 and

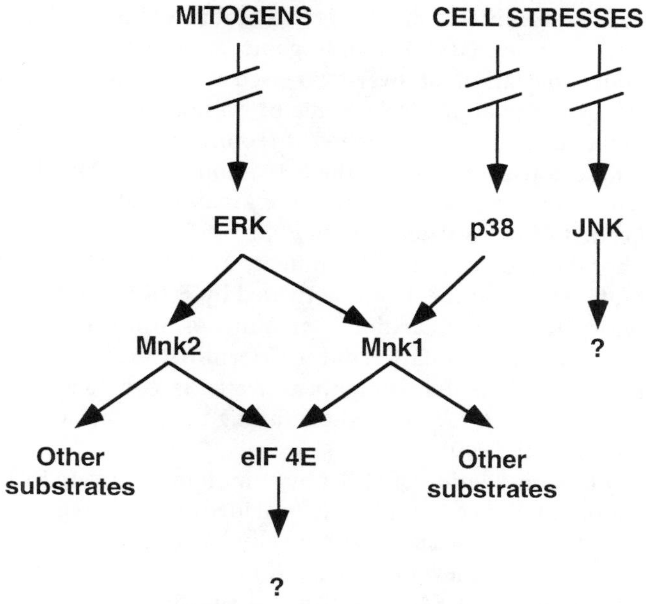

Fig. 1. Mnk1 and Mnk2 are closely related serine/threonine protein kinases that are activated by MAP kinase pathways and phosphorylate eIF4E at the physiologically relevant serine residue

Mnk2 bind these MAP kinases with different affinities. The CD region binds to basic regions on Mnk1 and Mnk2 that are close to the carboxy terminus, SRLARRRALA in Mnk1 and SKLAQRRQRA in Mnk2. Presumably, the differences between these two sequences are sufficient to confer the specific binding of p38 only to Mnk1, while ERK2 binds to both Mnk1 and Mnk2, and JNK1 binds to neither.

High-affinity binding of kinase to substrate is quite common, and is known as substrate docking (Holland and Cooper 1999). Potentially, high-affinity interactions outside the active site compensate for limited contacts between substrate and kinase in the active site, and help ensure specificity. In the case of Mnk1 and ERK, the complex is needed for efficient activation of Mnk1, since mutation of the docking site on Mnk1 inhibits activation (Tanoue et al. 2000). The complex may be preformed in resting cells and dissociate upon stimulation, because Mnk1 was found to preferentially associate with an underphosphorylated, inactive form of ERK2 (Waskiewicz et al. 1997).

Evidence that ERK2 actually phosphorylates Mnk1 was provided by Fukunaga and Hunter (1997) who cloned Mnk1 by virtue of its ability to serve as a substrate for ERK2. It can be phosphorylated in vitro by either ERK2 or p38. In cells it becomes more extensively phosphorylated, as evidenced by a mobility shift on an SDS-PAGE gel, after mitogen or stress treatments that activate ERKs and p38s, respectively. Using specific inhibitors of ERKs and p38s, Fukunaga and Hunter showed that this mobility shift of Mnk1 is dependent on these upstream kinases. The mitogenic stimulus, phorbol ester, stimulates the phosphorylation of over-expressed Mnk1 on several peptides in 293 cells (Waskiewicz et al. 1999). Some of these phosphopeptides were identified and found to correspond to two threonine residues that are contained in MAP kinase consensus sites in the activation loop of Mnk1. The activation loop is a region found in serine/threonine kinases that contains phosphorylation sites needed for activation (Zhang et al. 1995). Both threonines appear to be phosphorylated and a double mutant, Mnk1, in which both these residues are mutated to alanine, is not activated by ERK2 in vitro. A third potential phosphorylation site near the C-terminus of Mnk1 appears not to be phosphorylated. However, mutation of the C-terminal site to an aspartic acid residue gives rise to an allele that is highly activated in comparison to the wild type kinase and is a useful tool in understanding the biological function of Mnk1 (Waskiewicz et al. 1999).

Phosphorylation of Mnk1 by stimuli that activate ERK or p38 causes its activation. Evidence for this was provided by Fukunaga and Hunter (1997) using an "in gel" kinase assay with myelin basic protein as an in vitro substrate. Furthermore, Waskiewicz et al. (1997) had found that both Mnk1 and Mnk2 could phosphorylate eIF4E in vitro, and showed that phosphorylation of Mnk1 by ERK or p38 MAP kinases in vitro stimulated its activity. Using the double mutant of Mnk1, lacking the phosphorylation sites in the activation loop, it was also possible to show that in vivo, Mnk1 required phosphorylation in the activation loop to be able to phosphorylate co-transfected eIF4E in 293 cells (Waskiewicz et al. 1999). This double mutant of Mnk1 acts as a dominant negative allele with regards to eIF4E phosphorylation.

Therefore, Mnk1 and Mnk2 are protein kinases that associate with, are phosphorylated by, and are activated by MAP kinases, which are themselves activated by kinase cascades. Such sequential kinase activation reactions allow for a switch-like response of the cell to gradual changes in external stimuli (Ferrell 1996, 1998)

3
Mnk1 Binding Proteins

In searching for a role for Mnk1 in cells, a yeast two-hybrid screen was performed using Mnk1 as a bait in order to identify potential cellular substrates (Waskiewicz et al. 1999). Multiple clones of two molecules came out of this screen; a relative of importin-alpha and a form of the eukaryotic initiation factor eIF4G.

The relevance of the interaction with importin-alpha is as yet unknown. Importin-alpha plays a pivotal role in the classical nuclear protein import pathway, shuttling between the nucleus and the cytoplasm (Laskey et al. 1996; Moroianu 1997; Kohler et al. 1999). It binds to nuclear localization signal-bearing proteins and functions as an adapter to access the importin-beta-dependent import pathway (Laskey et al. 1996; Moroianu 1997; Kohler et al. 1999). However, since current antibodies to Mnk1 are not sufficiently sensitive to detect endogenous protein by immunofluorescence, it has been difficult to determine whether Mnk1 is imported into the nucleus. Immunofluorescence of overexpressed Mnk1 failed to show significant nuclear staining (Waskiewicz et al. 1997). It will be important to determine Mnk1 localization under various conditions of cell stimulation before excluding the possibility that Mnk1 may be nuclear under some conditions and that Mnk1 may be transported by importin-alpha. It is not known whether importin-alpha can be phosphorylated by Mnk1. Importin-alpha may yet be a cytoplasmic substrate for Mnk1, and Mnk1 may be important for the regulation of nuclear protein import.

4
Association With and Phosphorylation of the eIF4G Scaffold

The yeast two-hybrid screen with Mnk1 demonstrated that the carboxy terminus of eIF4GII bound to Mnk1 (Waskiewicz et al. 1999). Further studies have shown that the C termini of eIF4G family members eIF4GI, 4GII, and the translational inhibitor p97/NAT-1/DAP5 (novel translational repressor-1/death-associated protein-5) bind to the amino-terminus of Mnk1 (Pyronnet et al. 1999; Waskiewicz et al. 1999).

eIF4GI and eIF4GII are very similar to each other. In order from the N-terminus, they have binding sites for poly A-binding protein (PABP-1), eIF4E, the 40 S ribosome-binding factor eIF3, two sites for the RNA-dependent helicase eIF4A, and Mnk1 (Lamphear et al. 1995; Imataka and Sonenberg 1997; Imataka et al. 1998; Pyronnet et al. 1999; Waskiewicz et al. 1999; Fig. 2). eIF4GI and II are both cleaved by picornavirus 2 A protease, severing the PABP-1 and eIF4E binding sites from the remainder of the molecule. 2 A protease has been used extensively to demonstrate the importance of eIF4G for cap-dependent translation (Lamphear et al. 1995; Lamphear and Rhoads 1996). IRES (internal ribosome entry site)-driven translation is not inhibited under these conditions (Lamphear et al. 1995; Novoa et al. 1997) The role of p97/NAT-1/DAP5 is somewhat different. It corresponds approximately to the C terminal two thirds of eIF4GI or II, lacking binding sites for PABP-1 and eIF4E, but has a partial binding site for eIF4A. However, it still has the potential to bind eIF3 and Mnk. p97/NAT-1/DAP5 probably makes "dead" complexes by sequestering away eIF3 and Mnk from productive translation complexes and has been thought to have a role in fas and p53-induced apoptosis (Henis Korenblit et al. 2000). A caspase-cleaved apoptotic form of p97/NAT-1/DAP5 is thought to upregulate its own production from an IRES contained in its 5′ UTR. Overexpression was shown to inhibit both cap-dependent and cap-independent translation and this protein was found to accumulate in cells undergoing apoptosis.

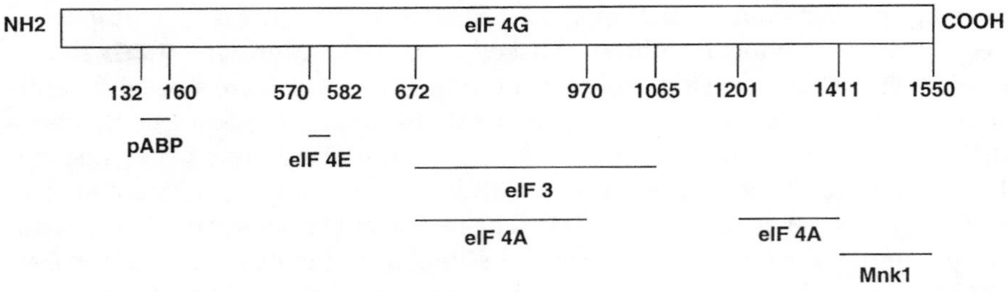

Fig. 2. eIF4G is a large scaffolding molecule that has binding sites for the indicated proteins. The binding regions have been specified by amino acid residue numbers

The potential effects of Mnk1 on p97/NAT-1/DAP5 function have not been explored.

eIF4GI appears to be a substrate for Mnk1, although the significance of this phosphorylation is as yet unknown (Pyronnet et al. 1999). eIF4GI is phosphorylated at the carboxy terminus in response to serum and mitogens (Raught et al. 2000). The sites of in vitro Mnk1 phosphorylation were mapped and are in a region between the first and second binding sites for eIF4A, which is poorly conserved in eIF4GII and p97/NAT-1/DAP5, neither of which have been reported to be phosphorylated by Mnk1. Surprisingly, phosphorylation of these sites was found to be repressed by serum stimulation in vivo. A deletion of the region containing the Mnk1 phosphorylation sites on eIF4G does not affect the phosphorylation of the serum-induced sites. It is possible that these same sites are upregulated during the stress response, as Mnk1 can be activated by p38 MAP kinase, but this has yet to be shown.

5 The eIF4E Kinase?

During the investigation of the significance of the interaction between eIF4G and Mnk1, eIF4G-associated proteins were tested as potential substrates for Mnk1. It was discovered that the cap-binding protein, eIF4E, is a substrate for Mnk1 and phosphorylation occurs at the physiologically relevant serine (Waskiewicz et al. 1997). eIF4E has been reported to be phosphorylated on a serine residue at position 209 in response to both mitogenic and some stress-activated signals (Flynn and Proud 1995; Morley 1997; Kleijn et al. 1998; Wang et al. 1998). The stimuli that induce phosphorylation of eIF4E closely parallel those that activate Mnk1; work has also shown that inhibition of ERK or p38 MAP kinase pathways can consequentially inhibit the phosphorylation of eIF4E (Wang et al. 1998). This correlation sets the precedent for Mnk1 being a eIF4E kinase in vivo. Furthermore, transient overexpression of an activated allele of Mnk1 in 293 cells increases the stoichiometry of phosphorylation of eIF4E, while a dominant-negative allele of Mnk1 inhibits the phosphorylation of eIF4E (Waskiewicz, et al. 1997). This provides more compelling evidence for Mnk1 or a Mnk1-like kinase to be an eIF4E kinase. However, it should be noted that casein kinase I, protein kinase C and an unknown protamine-activated protein kinase have also been reported to phosphorylate eIF4E in vitro, albeit not only at serine 209 (Haas and Hagedorn 1991, 1992; Amick and Damuni 1992; Makkinje et al. 1995). It is quite possible that kinases other than Mnk1 phosphorylate eIF4E in some cell types. The ability of the dominant-negative Mnk1 mutant to interfere with phosphorylation of eIF4E in 293 cells shows that a kinase with similar binding properties to Mnk1, that can be competed by the overexpressed protein, is responsible for the phosphorylation.

Pyronnet et al. (1999) took an alternative approach to characterize the eIF4E kinase. They showed that eIF4E needs its eIF4G binding site for basal phosphorylation in unstimulated 293T cells. This implies that the endogenous eIF4E kinase is associated with eIF4G, consistent with its identity with Mnk1. However, when overexpressed, Mnk1 seems to be able to phosphorylate eIF4E whether or not the eIF4E is associated with eIF4G (Waskiewicz et al. 1999). This was shown by overexpressing the 4EBPs that compete with eIF4G for a common binding site on eIF4E and, therefore, reduce the amount of eIF4E bound to eIF4G. Under these conditions, the eIF4E that was associated with the 4EBPs was fully phosphorylated by overexpressed Mnk1. Thus, when Mnk1 is in excess, it may phosphorylate eIF4E regardless of whether it is bound to eIF4G, but when only endogenous levels of Mnk1 are available, association of both eIF4E and Mnk1 with the same molecule of eIF4G may be necessary for efficient phosphorylation of eIF4E.

6
Possible Consequences of eIF4E Phosphorylation

The only known site of phosphorylation of eIF4E in cells is the Mnk-catalyzed site, serine 209. Phosphorylation at this site was originally detected in mitogen-stimulated cells and in transformed cells (Rychlik et al. 1987; Frederickson et al. 1992; Mendez et al. 1996; McKendrick et al. 1999). It seemed possible that phosphorylation of eIF4E may regulate recruitment of specific capped mRNAs to the ribosome. Messages with complex secondary structures in their 5′ UTR, some of which encode for growth regulatory gene products, are more dependent on eIF4E (Koromilas et al. 1992). The observations that malignant transformation and growth factor-independence could be caused by eIF4E overexpression, and not by overexpression of a mutant of eIF4E that is not phosphorylated (De Benedetti and Rhoads 1990; Lazaris Karatzas et al. 1992; Lazaris Karatzas and Sonenberg 1992) lent support to a model that eIF4E levels and phosphorylation were important for cell growth. However, while eIF4E availability is clearly regulated during the mitogenic response, by release from 4EBPs, the relevance of eIF4E phosphorylation has remained unclear.

A number of developments have undermined the model. First, it was found that the phosphorylation site in eIF4E had been misidentified (Flynn and Proud 1995). The non-phosphorylated mutant of eIF4E used as a control in the transformation experiments was mutated at a serine that appears not to be phosphorylated, and its lack of function may be due to an altered structure that precludes phosphorylation of the correct site, serine 209. The transforming activity of eIF4E mutants with serine 209 substituted has not yet been reported. Second, it has become clear that, even though phosphorylation occurs when translation is increasing and many cell mRNAs are recruited into larger polysomes, phosphorylation also occurs under conditions where translation is reduced, such as in cells stressed with anisomycin, arsenite and heat shock (Morley 1997a, b; Morley and McKendrick 1997; Wang et al. 1998). In addition, eIF4E phosphorylation is reduced, but translation increases, when L6 myoblasts are treated with insulin (Kimball et al. 1998). In Ha-ras transformed rat embryo fibroblasts, there is no change in the stoichiometry of eIF4E phosphorylation, even though there is a sevenfold increase in both phosphorylation and dephosphorylation rates (Rinker Schaeffer et al. 1992), suggesting that phosphate turnover might, in fact, be an important factor in regulating eIF4E function. Thus, there is poor correlation between eIF4E phosphorylation state and overall protein synthesis rate. Third, transformed cells that are stably overexpressing eIF4E show increased translation of mRNAs with 5′ UTR secondary structure (Manzella and Blackshear 1990; Manzella et al. 1991; Haghighat et al. 1995), but this could be a secondary effect of the transformed phenotype rather than a primary consequence of eIF4E overexpression. Indeed, transient overexpression of eIF4E was found not to correlate with increases in synthesis of total protein or specific reporters (Kaufman et al. 1993). Therefore, it remains uncertain whether eIF4E phosphorylation regulates protein synthesis in cells.

Nonetheless, phosphorylation of eIF4E does appear to regulate binding to capped mRNA (Lamphear and Panniers 1990; Minich et al. 1994). Minich et al. (1994) obtained phosphorylated and unphosphorylated eIF4E from reticulocyte extracts, and showed that the former has three- to fourfold greater affinity for globin mRNA than the latter. This affinity was reduced by treating the phosphorylated eIF4E with phosphatase. The phosphorylation site is close to the groove that accommodates the cap structure, and adding the negative phosphate group has been proposed to form a salt bridge with a lysine on the other side of the groove, thus trapping previously bound cap in place (Marcotrigiano et al. 1997a, b). Depending on when the phosphorylation occurs, however, it could equally be argued that prior phosphorylation might prevent subsequent capped mRNA binding or inhibit the release of the mRNA that is already bound to eIF4E.

There may also be unsuspected changes in mRNA binding due to interactions between eIF4E and other proteins. 4EBPs and eIF4G compete for the same binding sites on eIF4E (Haghighat et al. 1995; Mader et al. 1995), on the opposite site of eIF4E from the cap (Matsuo et al. 1997) (Fig. 3). Haghighat and

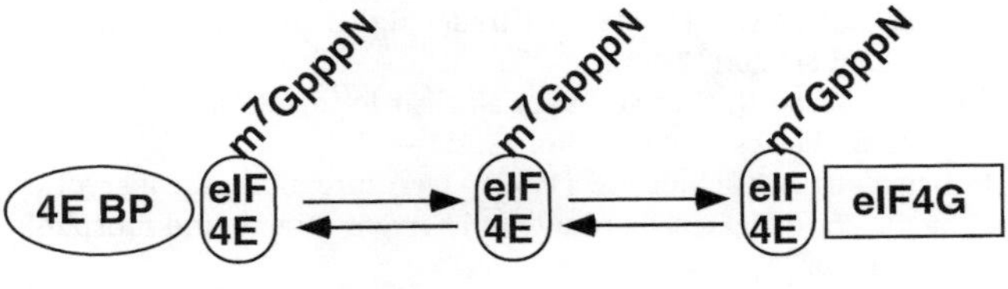

Fig. 3. The levels of eIF4E in cells are regulated by virtue of its association to a family of binding proteins called 4E binding proteins (4EBP). 4EBPs compete for binding to the same sites on 4E as does eIF4G. The cap structure m^7GpppN binds to eIF4E, possibly with differing affinities when it is free or bound to associated proteins

Sonenberg (1997) and Ptushkina et al. (1999) have reported that binding of eIF4E to eIF4G and 4EBPs increases the affinity for capped mRNA. In that case, some of the change in affinity detected by Minich et al. (1994) may have been due to binding proteins associated with the eIF4E in the cell extracts. The discovery that the eIF4G/4EBP-binding site regulates the cap-binding site on eIF4E also raises the possibility that serine 209 phosphorylation could reciprocally affect the binding of eIF4E to other proteins.

7 Conclusion

There are, thus, many unanswered questions regarding the functional consequences of eIF4E phosphorylation at serine 209. Studies in lower organisms have not resulted in any information regarding the phosphorylation of eIF4E at serine 209. A Mnk1 homologue has been found in *X. laevis* (78.5% amino acid identity; GenBank accession no. AB023807), where it may be involved in the eIF4E phosphorylation that occurs during meiotic maturation of oocytes (Morley and Pain 1995). Further studies in this system may help to resolve the importance of phosphorylation of eIF4E in translation regulation.

Unraveling the significance of eIF4E phosphorylation at serine 209 will require in vitro biochemistry and in vivo approaches, for which mutant alleles of Mnk1 may prove useful. In addition, the structure of phosphorylated eIF4E may help to resolve the impending issues and is eagerly awaited.

Acknowledgements. The authors wish to thank Bennett Penn, Leslie Cary and Andrew Waskiewicz for critical reading of the chapter, and acknowledge the NIH grant RO1-CA-73897 (to J.A.C.) for support.

References

Ahn NG, Seger R et al (1992) Growth factor-stimulated phosphorylation cascades: activation of growth factor-stimulated MAP kinase. Ciba Found Symp 164:113–126

Amick GD, Damuni Z (1992) Protamine kinase phosphorylates eukaryotic protein synthesis initiation factor 4E. Biochem Biophys Res Commun 183:431–437

Cano E, Mahadevan LC (1995) Parallel signal processing among mammalian MAPKs. Trends Biochem Sci 20:117–122

Chen RH, Sarnecki C et al (1992) Nuclear localization and regulation of erk- and rsk-encoded protein kinases. Mol Cell Biol 12:915–927

De Benedetti A, Rhoads RE (1990) Overexpression of eukaryotic protein synthesis initiation factor 4E in HeLa cells results in aberrant growth and morphology. Proc Natl Acad Sci USA 87:8212–8216

Duncan R, Milburn SC et al (1987) Regulated phosphorylation and low abundance of HeLa cell initiation factor eIF-4F suggest a role in translational control Heat shock effects on eIF-4F. J Biol Chem 262:380–388

Ferrell JJ (1996) Tripping the switch fantastic: how a protein kinase cascade can convert graded inputs into switch-like outputs. Trends Biochem Sci 21:460–466

Ferrell JJ (1998) How regulated protein translocation can produce switch-like responses. Trends Biochem Sci 23:461–465

Flynn A, Proud CG (1995) Serine 209, not serine 53, is the major site of phosphorylation in initiation factor eIF-4E in serum-treated Chinese hamster ovary cells. J Biol Chem 270:21684–21688

Frederickson RM, Montine KS et al (1991) Phosphorylation of eukaryotic translation initiation factor 4E is increased in Src-transformed cell lines. Mol Cell Biol 11:2896–2900

Frederickson RM, Mushynski WE et al (1992) Phosphorylation of translation initiation factor eIF-4E is induced in a ras-dependent manner during nerve growth factor-mediated PC12 cell differentiation. Mol Cell Biol 12:1239–1247

Fukunaga R, Hunter T (1997) MNK1, a new MAP kinase-activated protein kinase, isolated by a novel expression screening method for identifying protein kinase substrates. EMBO J 16:1921–1933

Graff JR, Boghaert ER et al (1995) Reduction of translation initiation factor 4E decreases the malignancy of ras-transformed cloned rat embryo fibroblasts. Int J Cancer 60:255–263

Graves LM, Guy HI et al (2000) Regulation of carbamoyl phosphate synthetase by MAP kinase. Nature 403:328–332

Haas DW, Hagedorn CH (1991) Casein kinase I phosphorylates the 25-kDa mRNA cap-binding protein. Arch Biochem Biophys 284:84–89

Haas DW, Hagedorn CH (1992) Protein kinase C phosphorylates both serine and threonine residues of the mRNA cap binding protein eIF-4E. Second Messengers Phosphoproteins 14:55–63

Haghighat A, Mader S et al (1995) Repression of cap-dependent translation by 4E-binding protein 1: competition with p220 for binding to eukaryotic initiation factor-4E. EMBO J 14:5701–5709

Haghighat A, Sonenberg N (1997) eIF4G dramatically enhances the binding of eIF4E to the mRNA 5′-cap structure. J Biol Chem 272:21677–21680

Henis-Korenblit S, Strumpf NL et al (2000) A novel form of DAP5 protein accumulates in apoptotic cells as a result of caspase cleavage and internal ribosome entry site-mediated translation. Mol Cell Biol 20:496–506

Hirabayashi T, Kume K et al (1998) Conditional expression of the dual-specificity phosphatase PYST1/MKP-3 inhibits phosphorylation of cytosolic phospholipase A2 in Chinese hamster ovary cells. Biochem Biophys Res Commun 253:485–488

Holland PM, Cooper JA (1999) Protein modification: docking sites for kinases. Curr Biol 9:R329–R331

Ichijo H (1999) From receptors to stress-activated MAP kinases. Oncogene 18:6087–6093

Imataka H, Gradi A et al (1998) A newly identified N-terminal amino acid sequence of human eIF4G binds poly(A)-binding protein and functions in poly(A)-dependent translation. EMBO J 17:7480–7489

Imataka H, Sonenberg N (1997) Human eukaryotic translation initiation factor 4G (eIF4G) possesses two separate and independent binding sites for eIF4 A. Mol Cell Biol 17:6940–6947

Kaufman RJ, Murtha RP et al (1993) Characterization of wild-type and Ser53 mutant eukaryotic initiation factor 4E overexpression in mammalian cells. J Biol Chem 268:11902–11909

Kimball SR, Horetsky RL et al (1998) Signal transduction pathways involved in the regulation of protein synthesis by insulin in L6 myoblasts. Am J Physiol 274:C221–C228

Kleijn M, Scheper GC et al (1998) Regulation of translation initiation factors by signal transduction. Eur J Biochem 253:531–544

Kohler M, Haller H et al (1999) Nuclear protein transport pathways Exp Nephrol 7:290–294

Koromilas AE, Lazaris Karatzas A et al (1992) mRNAs containing extensive secondary structure in their 5′ non-coding region translate efficiently in cells overexpressing initiation factor eIF-4E. EMBO J 11:4153–4158

Lamphear BJ, Kirchweger R et al (1995) Mapping of functional domains in eukaryotic protein synthesis initiation factor 4G (eIF4G) with picornaviral proteases. Implications for cap-dependent and cap-independent translational initiation. J Biol Chem 270:21975–21983

Lamphear BJ, Panniers R (1990) Cap binding protein complex that restores protein synthesis in heat-shocked Ehrlich cell lysates contains highly phosphorylated eIF-4E. J Biol Chem 265:5333–5336

Lamphear BJ, Rhoads RE (1996) A single amino acid change in protein synthesis initiation factor 4G renders cap-dependent translation resistant to picornaviral 2 A proteases. Biochemistry 35:15726–15733

Laskey RA, Gorlich D et al (1996) Regulatory roles of the nuclear envelope. Exp Cell Res 229:204–211

Lazaris Karatzas A, Montine KS et al (1990) Malignant transformation by a eukaryotic initiation factor subunit that binds to mRNA 5′ cap. Nature 345:544–547

Lazaris Karatzas A, Smith MR et al (1992) Ras mediates translation initiation factor 4E-induced malignant transformation. Genes Dev 6:1631–1642

Lazaris Karatzas A, Sonenberg N (1992) The mRNA 5′ cap-binding protein, eIF-4E, cooperates with v-myc or E1 A in the transformation of primary rodent fibroblasts. Mol Cell Biol 12:1234–1238

Mader S, Lee H et al (1995) The translation initiation factor eIF-4E binds to a common motif shared by the translation factor eIF-4 gamma and the translational repressors 4E-binding proteins. Mol Cell Biol 15:4990–4997

Mader S, Sonenberg N (1995) Cap binding complexes and cellular growth control. Biochimie 77:40–44

Makkinje A, Xiong H et al (1995) Phosphorylation of eukaryotic protein synthesis initiation factor 4E by insulin-stimulated protamine kinase. J Biol Chem 270:14824–14828

Manzella JM, Blackshear PJ (1990) Regulation of rat ornithine decarboxylase mRNA translation by its 5′-untranslated region. J Biol Chem 265:11817–11822

Manzella JM, Rychlik W et al (1991) Insulin induction of ornithine decarboxylase Importance of mRNA secondary structure and phosphorylation of eucaryotic initiation factors eIF-4B and eIF-4E. J Biol Chem 266:2383–2389

Marcotrigiano J, Gingras A-C et al (1997a) Cocrystal structure of the messenger RNA 5′ cap-binding protein (eIF4E) bound to 7-methyl-GDP. Cell 89:951–961

Marcotrigiano J, Gingras A-C et al (1997b) X-ray studies of the messenger RNA 5′ cap-binding protein (eIF4E) bound to 7-methyl-GDP. Nucleic Acids Symp Ser 36:8–11

Matsuo H, Li H et al (1997) Structure of translation factor eIF4E bound to m7GDP and interaction with 4E-binding protein. Nat Struct Biol 4:717–724

McKendrick L, Pain VM et al (1999) Translation initiation factor 4E. Int J Biochem Cell Biol 31:31–35

Mendez R, Myers MJ et al (1996) Stimulation of protein synthesis, eukaryotic translation initiation factor 4E phosphorylation, and PHAS-I phosphorylation by insulin requires insulin receptor substrate 1 and phosphatidylinositol 3-kinase. Mol Cell Biol 16:2857–2864

Minich WB, Balasta ML et al (1994) Chromatographic resolution of in vivo phosphorylated and nonphosphorylated eukaryotic translation initiation factor eIF-4E:increased cap affinity of the phosphorylated form. Proc Natl Acad Sci USA 91:7668–7672

Morley SJ (1997a) Intracellular signalling pathways regulating initiation factor eIF4E phosphorylation during the activation of cell growth. Biochem Soc Trans 25:503–509

Morley SJ (1997b) Signalling through either the p38 or ERK mitogen-activated protein (MAP) kinase pathway is obligatory for phorbol ester and T cell receptor complex (TCR-CD3)-stimulated phosphorylation of initiation factor (eIF) 4E in Jurkat T cells. FEBS Lett 418:327–332

Morley SJ, McKendrick L (1997) Involvement of stress-activated protein kinase and p38/RK mitogen-activated protein kinase signaling pathways in the enhanced phosphorylation of initiation factor 4E in NIH 3T3 cells. J Biol Chem 272:17887–17893

Morley SJ, Pain VM (1995) Hormone-induced meiotic maturation in Xenopus oocytes occurs independently of p70s6k activation and is associated with enhanced initiation factor (eIF)-4F phosphorylation and complex formation. J Cell Sci 108:1751–1760

Moroianu J (1997) Molecular mechanisms of nuclear protein transport. Crit Rev Eukaryot Gene Exp 7:61–72

Nishida E, Gotoh Y (1993) The MAP kinase cascade is essential for diverse signal transduction pathways. Trends Biochem Sci 18:128–131

Novoa I, Martinez AF et al (1997) Cleavage of p220 by purified poliovirus 2 A(pro) in cell-free systems: effects on translation of capped and uncapped mRNAs. Biochemistry 36:7802–7809

Ptushkina M von et al (1999) Repressor binding to a dorsal regulatory site traps human eIF4E in a high cap-affinity state. EMBO J 18:4068–4075

Pyronnet S, Imataka H et al (1999) Human eukaryotic translation initiation factor 4G (eIF4G) recruits mnk1 to phosphorylate eIF4E. EMBO J 18:270–279

Raught B, Gingras A-C et al (2000) Serum-stimulated, rapamycin-sensitive phosphorylation sites in the eukaryotic translation initiation factor 4GI. EMBO J 19:434–444

Rinker Schaeffer CW, Austin CV et al (1992) Ras transformation of cloned rat embryo fibroblasts results in increased rates of protein synthesis and phosphorylation of eukaryotic initiation factor 4E. J Biol Chem 267:10659–10664

Robbin DJ, Cheng M et al (1992) Evidence for a Ras-dependent extracellular signal-regulated protein kinase (ERK) cascade. Proc Natl Acad Sci USA 89:6924–6928

Rosenwald IB, Rhoads DB et al (1993) Increased expression of eukaryotic translation initiation factors eIF-4E and eIF-2 alpha in response to growth induction by c-myc. Proc Natl Acad Sci USA 90:6175–6178

Rychlik W, Russ MA et al (1987) Phosphorylation site of eukaryotic initiation factor 4E. J Biol Chem 262:10434–10437

Tanoue T, Adachi M et al (2000) A conserved docking motif in MAP kinases common to substrates, activators and regulators. Nat Cell Biol 2:110–116

Wang X, Flynn A et al (1998) The phosphorylation of eukaryotic initiation factor eIF4E in response to phorbol esters, cell stresses, and cytokines is mediated by distinct MAP kinase pathways. J Biol Chem 273:9373–9377

Waskiewicz AJ, Flynn A et al (1997) Mitogen-activated protein kinases activate the serine/threonine kinases Mnk1 and Mnk2. EMBO J 16:1909–1920

Waskiewicz AJ, Johnson JC et al (1999) Phosphorylation of the cap-binding protein eukaryotic translation initiation factor 4E by protein kinase Mnk1 in vivo. Mol Cell Biol 19:1871–1880

Xu XX, Rock CO et al (1994) Regulation of cytosolic phospholipase A2 phosphorylation and eicosanoid production by colony-stimulating factor 1. J Biol Chem 269:31693–31700

Zhang J, Zhang F et al (1995) Activity of the MAP kinase ERK2 is controlled by a flexible surface loop. Structure 3:299–307

Control of Translation by the Target of Rapamycin Proteins

Anne-Claude Gingras, Brian Raught, and Nahum Sonenberg[1]

1 Introduction

Regulation of translation rates, the frequency with which a given mRNA is translated, plays an important role in the control of cell growth and differentiation. Translational control is exerted in most instances at the initiation phase, a rate-limiting step during which the ribosome is recruited to mRNA. Initiation is a complex process mediated by many translation initiation factors (at least 30 polypeptides), and the regulation of translation initiation factor activity involves modulation of gene expression, binding to other factors or repressors, proteolytic cleavage and changes in phosphorylation state. It has been known for some time that the phosphorylation state of various translation factors/inhibitors (and other proteins required for translation, such as ribosomal proteins) is modulated in response to hormonal/mitogenic signals and environmental or nutritional stresses, but the identity of the signaling pathways involved in translational regulation are only beginning to emerge. In this review, we describe a signaling module involved in translational control both in yeast and in mammalian cells, the TOR (or FRAP/mTOR) signaling pathway. In mammals, this pathway regulates the activity of several translation factors (eIF4B and eIF4GI), translation inhibitors (the 4E-BPs), and the ribosomal S6 kinases (S6K1 and 2). In yeast, inhibition of Tor activity leads to polysomal disaggregation and G1 cell cycle arrest. Disruption, mutation, or amplification of several genes leads to a partial or complete rapamycin-resistance, which has enabled the identification of several important players in this pathway. Inhibition of Tor activity also signals to translation initiation in yeast through degradation of the eIF4G proteins. Interestingly, the TOR pathway was also shown in *S. cerevisiae* to control the transcription of ribosomal proteins, ribosomal RNAs, and mRNAs coding for several translation initiation factors. In addition, Tor proteins have been recently shown to function in a nutrient-sensing pathway, which induces or represses transcription of several metabolic enzymes. The characterization of this pathway began about 25 years ago, after the discovery of the antifungal compound rapamycin.

[1] Department of Biochemistry and McGill Cancer Centre, McGill University, 3655 Dummond Street, Montréal, Québec, H3G 1Y6, Canada

Progress in Molecular and Subcellular Biology, Vol. 27
Robert E. Rhoads (Ed.)

2
Rapamycin (Also Known As Sirolimus or Rapamune)

2.1
Discovery and Activity of Rapamycin

The lipophilic macrolide rapamycin was isolated from a strain of *Streptomyces hygroscopicus* found in a soil sample from Easter Island (known by the inhabitants as Rapa Nui; Vezina et al. 1975). This compound was initially observed to be a potent inhibitor of yeast growth, being most effective against *Candida albicans* (Baker et al. 1978; Sehgal et al. 1975; Singh et al. 1979). Rapamycin is without effect on bacterial growth, and inhibits the proliferation of different eukaryotic microorganisms to various degrees. Early evidence that rapamycin also acts as an immunosuppressant in higher organisms came from its preventive effects on two experimental immunopathies in rat models (experimental autoimmune encephalitis and adjuvant arthritis; Martel et al. 1977). In the same study, rapamycin was also demonstrated to prevent formation of humoral IgE antibody. A decade passed before the mechanism of immunosuppression by rapamycin was elucidated: T cell proliferation, when induced by stimulation of the CD3/TCR or the CD28 pathways, is significantly inhibited by rapamycin. Rapamycin also inhibits the proliferation of activated T cells when stimulated further by IL-2, IL-4 or phorbol esters (Bierer et al. 1991; Dumont et al. 1990a, b; Kay et al. 1991; Luo et al. 1993). Rapamycin also potently inhibits the proliferation of several other lymphoid and non-lymphoid cell types, and the growth of several tumor cell lines (e.g., Albers et al. 1993; Eng et al. 1984; Kay et al. 1991). Interestingly, however, the extent of inhibition of proliferation between cell types and/or cell lines varies, even amongst cell lines of the same origin. For example, the BJAB B-lymphoblastoid cell line is much more sensitive to rapamycin than other B-lymphoblastoid cell lines (Kay et al. 1996). Because all cell lines tested display some degree of sensitivity to rapamycin, it was postulated that the "rapamycin receptor" must have a broad expression pattern.

The immunosuppressive properties of rapamycin have important ramifications for human health. Studies with animal transplantation models revealed that rapamycin is potent at preventing allograft rejection (reviewed in Abraham and Wiederrecht 1996). Clinical trials performed on 700 organ transplant patients (principally kidney transplants) receiving a combination of rapamycin, cyclosporin and corticosteroids revealed that rapamycin reduces acute rejection rates by up to 60%. Thus, rapamycin is used routinely in the clinic today, in combination with other immunosuppressants (reported by Bradley 1999).

2.2
Rapamycin-Related Compound FK506

The characterization of the molecular mechanism of action of rapamycin was stimulated by the discovery of a related compound termed FK506, isolated from *Streptomyces tsukabaensis* found in a Japanese soil sample (Kino et al. 1987a). FK506 was also demonstrated to be a potent immunosuppressant, as measured by its effect in many immune function assays (Kino et al. 1987a, b). FK506 is structurally similar to rapamycin (Fig. 1), and a portion of the molecule (as indicated by the shaded area) is nearly identical. However, while both

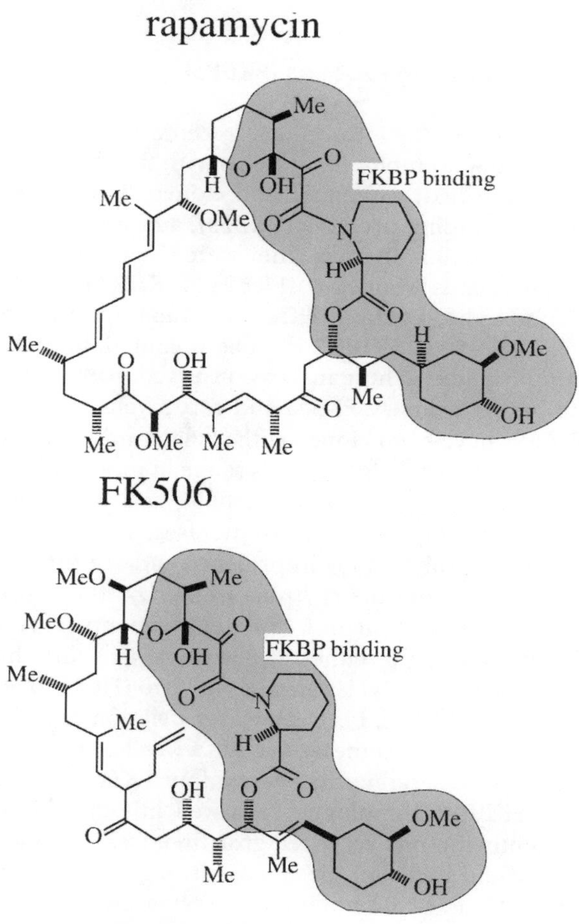

Fig. 1. Chemical structures of rapamycin and FK506. The chemical structures of the portion of the molecules shown to bind to the FKBPs is nearly identical (*shaded area*). However, the remainder of the molecules (the effector domain) differs. The effector domains are responsible for the different cellular responses elicited by the two compounds

rapamycin and FK506 possess immunosuppressive properties, it was soon observed that the two compounds do not inhibit the same intracellular signaling pathways. FK506 prevents transmission of signals from the T cell antigen receptor (TCR) via inhibition of calcineurin (or PP2B), a Ca^{2+}-regulated serine-threonine phosphatase (reviewed in Abraham and Wiederrecht 1996). Rapamycin is without effect on PP2B, but interferes with a pathway leading to cell cycle progression (G1-S), as discussed below. Rapamycin and FK506 are mutually antagonistic (Bierer et al. 1990; Dumont et al. 1990a, b).

3 Immunophilins

3.1 FK506-Binding Proteins (FKBPs)

To identify the intracellular receptor(s) for FK506, an affinity purification method was employed. Proteins with the ability to bind to a FK506 solid matrix were purified, sequenced and cloned. The molecules identified were termed FK506-binding proteins (FKBPs), and are also more generally referred to as immunophilins, because they were identified as proteins that bind to immunosuppressants (see below). FKBP12 (FKBP, MW of 12 kDa) was purified first, and found to bind to both FK506 and rapamycin (Harding et al. 1989; Siekierka et al. 1989, 1990). FKBP12 is an abundant, ubiquitously expressed protein of 108 amino acids (in humans), is conserved from yeast to mammals, and is thought to be the primary FK506 and rapamycin receptor. Subsequently, many other FKBPs have been cloned, either via affinity for FK506 or by sequence homology to known FKBPs. FKBPs range in molecular weight from 12 to 52 kDa, and exhibit varying patterns of cellular and subcellular expression. All FKBPs are peptidyl-prolyl *cis/trans* isomerases, which catalyze the interconversion of peptidyl-prolyl bonds in proteins, and are presumed to play a role in protein folding because the *cis/trans* isomerization of peptidyl-prolyl bonds is limiting in in vitro protein folding experiments (reviewed in Gothel and Marahiel 1999). Binding of either FK506 or rapamycin inhibits the catalytic activity of the peptidyl-prolyl isomerases in vitro (Heitman et al. 1991b; Koltin et al. 1991; Wiederrecht et al. 1991). However, deletion experiments in yeast suggested that the loss of the isomerase activity was not the cause of rapamycin (or FK506) sensitivity. *S. cerevisiae* possess four *FKBP* genes (termed *FPR1–4*, *FPR1* being the FKBP12 homologue; reviewed in Dolinski et al. 1997). Since rapamycin potently inhibits yeast cell growth, it was postulated that deletion of its receptor would be lethal (or at least strongly impair growth). However, ablation of the *FPR1* gene did not influence cell growth. Rather, the loss of *FPR1* conferred recessive rapamycin-resistance, indicating that the presence of Fpr1p (but not its catalytic activity) was required for rapamycin action (Heitman et al. 1991b; Koltin et al. 1991; Wiederrecht et al. 1991). Individual deletion of the other *FPR* genes was also not lethal (e.g., Dolinski et al. 1997). In fact, yeast remained viable even after deletion of all four *FPR* genes (Dolinski et al. 1997), arguing against a compensatory mechanism for the loss of the Fpr1p protein.

3.2 Cyclophilins and Parvulins

The mechanism of action of another immunosuppressant (which bears no structural homology to FK506 and rapamycin), cyclosporin A (CsA), was also investigated using a similar affinity purification scheme. Proteins that bind to CsA (or which possess sequence homology to previously cloned CsA binding proteins) have been collectively termed cyclophilins. Eight *S. cerevisiae* cyclophilins have been cloned, all of which are prolyl isomerases (reviewed in Gothel and Marahiel 1999). No sequence homology exists between the cyclophilins and the FKBPs. CsA selectively binds to and inhibits the isomerase activity of cyclophilins, without affecting that of the FKBPs. Similar to the deletion analysis of *FPR* genes, deletion of *CPR1* does not alter yeast growth or viability, but instead confers recessive resistance to CsA treatment (Haendler et al. 1989; Tropschug et al. 1989; Sykes et al. 1993). Individual deletion of each of the eight *CPR* genes does not alter viability. Combined deletion of all eight *CPR* genes also yields viable cells, and combined disruption of all eight *CPR* genes and all four *FPR* genes does not result in a loss of viability (Dolinski et al. 1997).

A third class of prolyl isomerases, termed the parvulins (from the Latin *parvus*, small), comprise a small family in eukaryotic cells. Unlike FKBPs and the cyclophilins, the parvulins are essential to cell cycle progression (reviewed in Gothel and Marahiel 1999). For example, depletion of *Pin1* (a mammalian parvulin) or *Ess1p* (its *S. pombe* homologue) induces mitotic arrest, and *Pin1* overexpression induces a G2 arrest. The mechanism of action of *Pin1* is thought to be mediated through the specific binding and isomerization of proteins phosphorylated on Ser/Thr-Pro sites, which could alter their conformation, and activity (reviewed in Lu 2000).

In summary, the FKBPs belong to a large family of proteins, the immunophilins, with a common enzymatic function, but with divergent biological roles. The precise function of most immunophilins remains elusive.

4 Identification of Rapamycin Targets

4.1 Cloning of Yeast Target of Rapamycin (TOR) Proteins

Since rapamycin does not function in vivo through an inhibition of the enzymatic activity of the FKBPs, a search for the authentic rapamycin target was initiated in yeast, using a genetic approach. The growth of *S. cerevisiae* is acutely sensitive to rapamycin, allowing for the selection of rapamycin-resistant mutants. This strategy led to the confirmation of Fpr1p as a rapamycin target, but also to the identification of two novel genes, termed *TOR1* and *2* (Target Of Rapamycin 1 and 2; also known as *DRR1* and *2*, for Dominant Rapamycin Resistant). Unlike the FKBPs, the TOR mutants (which are point mutants, as described in Sect. 5.1) act in a dominant (or semi-dominant) manner (Heitman et al. 1991a; Cafferkey et al. 1993; Kunz et al. 1993; Helliwell et al. 1994; Lorenz and Heitman 1995). Tor1p and Tor2p are large proteins (>280 kDa), sharing 67% identity at the amino acid level. Because their C-terminal region contains a conserved domain with homology to lipid kinases (Fig. 2), they have been placed in a family of signaling molecules termed the PIKKs (for phosphoinositide kinase-related kinases; see Sect. 5.2). Tor1p and Tor2p were demonstrated to form complexes with Fpr1p-rapamycin (but not with Fpr1p or rapamycin alone), validating the model by which Fpr1p-rapamycin forms a gain-of-function complex.

Disruption of only the *TOR1* gene is not lethal, but leads to a slight decrease in cell growth (Helliwell et al. 1994). When treated with rapamycin, *TOR1* deletants arrest in G1 within one generation, and rapamycin-sensitivity is increased four-fold as compared to the wild type strain (Helliwell et al. 1994; Lorenz and Heitman 1995) In a *TOR1* deleted strain, overexpression of *TOR2* decreased rapamycin-sensitivity, indicating that the Tor proteins are limiting

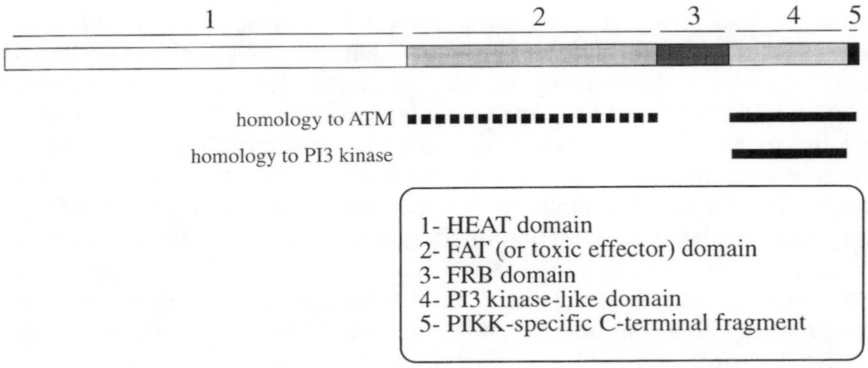

Fig. 2. Modular structure of FRAP/mTOR. Five structural elements have been described in the FRAP/mTOR sequence (labeled *1–5*, from the N-terminus to the C-terminus). *1* The HEAT domain is comprised of repeats of 40 amino acids and spans roughly the first 1200 amino acids of FRAP/mTOR. This portion of the molecule was documented to bind gephyrin. *2* The FAT domain (a.k.a. the toxic effector domain) is toxic for yeast growth: this domain is conserved amongst PIKK and TRAPP proteins. *3* The FRB domain is responsible for the binding of the FKBP12-rapamycin complex; a point mutant in this domain (at Ser2035) generates a dominant rapamycin-resistant FRAP/mTOR. *4* The kinase domain of FRAP/mTOR is most closely related to that of the other PIKKs, but also to lipid kinases, such as PI3 kinases and PI4 kinases. The kinase domain is essential to FRAP/mTOR activity. *5* The C-terminal tail is extremely conserved amongst the PIKKs, but is not present in PI3 kinases or PI4 kinases. The function of this domain is unknown. Regions bearing homology to the other PIKKs (here ATM) and to PI3 kinases are identified. The *dashed line* indicates a lesser degree of homology. Please refer to text for details

for FKBP-rapamycin action (Lorenz and Heitman 1995). Disruption of *TOR2* is lethal, but does not cause a cell-cycle phase specific arrest (Kunz et al. 1993). Instead, *TOR2* disrupted cells arrest randomly throughout the cell cycle. Strains disrupted for both *TOR1* and *TOR2* recapitulate the phenotype observed with rapamycin treatment, in which cells arrest in G1 within one generation (Helliwell et al. 1994; Kunz et al. 1993). Taken together, these data indicate that the presence of either Tor1p or Tor2p is sufficient for G1 progression, and that Tor function (provided by either protein) is necessary for G1 progression. This Tor G1 activity is sensitive to inhibition by rapamycin, and inhibition of both proteins is required for the cell cycle arrest effected by rapamycin.

The lethality of the *TOR2* deletion is not restored by overexpression of wild-type or rapamycin-resistant *TOR1* (Zheng et al. 1995). This lack of complementation indicates that Tor2p, in addition to a shared rapamycin-inhibitable activity with Tor1p, also possesses a rapamycin-resistant activity which is essential but not shared with Tor1p (Kunz et al. 1993; Zheng et al. 1995). The Tor2p-specific function appears to be linked to cell-cycle-dependent polarization of the actin cytoskeleton, which is disrupted in a *tor2* mutant (Schmidt et al. 1996). Signaling from Tor proteins to the cytoskeleton has been shown to involve the activity of the Rho-like GTPases Rho1p and Rho2p, or their exchange factor Rom2p, and the Rho1p GTPase activating enzyme Sac7p (Schmidt et al. 1997). This Tor2p-specific function does not reside in the catalytic domain, as these are interchangeable between Tor1p and Tor2p (Helliwell et al. 1994).

4.2 Cloning of the Mammalian Target of Rapamycin Proteins

The mammalian homologue of the Tor proteins was cloned from several different species a few years after the yeast proteins were isolated. The strategy utilized for the purification and identification of the mammalian Tor homologues was, in most cases, affinity purification based on the principle that the protein should bind a complex of rapamycin and FKBP12. Large molecular weight proteins were purified from various sources, and alternatively termed FRAP (FKBP and rapamycin-associated protein), mTOR (mammalian target of rapamycin), RAFT (rapamycin and FKBP12 target), SEP (sirolimus effector protein) and RAPT (rapamycin target; Brown et al. 1994; Chen et al. 1994; Chiu et al. 1994; Sabatini et al. 1994; Sabers et al. 1995). For clarity, here we refer to the mammalian protein as FRAP/mTOR. Initial validation of FRAP/mTOR as the mammalian target of rapamycin was provided by studies with rapamycin analogs and rapamycin-resistant cell lines. FRAP/mTOR bound poorly (if at all) to two structural analogs of rapamycin (25,26 iso-rapamycin and 16-keto-rapamycin), which bind FKBP12 with high affinity, but which are poor inhibitors of G1 progression (Brown et al. 1994). Furthermore, in murine T cell (YAC) lines selected for rapamycin-resistance (Dumont et al. 1995), little or no

FRAP/mTOR associates with FKBP12-rapamycin, while wild-type levels of association are found in a rapamycin-sensitive YAC revertant (Sabers et al. 1995). FRAP/mTOR is 289 kDa and shares roughly 45% identity with the yeast Tor proteins (Brown et al. 1994; Sabatini et al. 1994; Sabers et al. 1995). While *S. cerevisiae* possesses two Tor proteins, only one homologue has been found in mammals. The human, rat and mouse (GenBank accession nos.: NP 004949, A54837 and AAF73196, respectively) FRAP/mTOR proteins are very similar, sharing more than 95% identity at the amino acid level. As with the yeast proteins, the cloned FRAP/mTOR was shown to bind in vitro to FKBP12-rapamycin (Sabers et al. 1995). FRAP/mTOR mRNA is ubiquitously expressed, with higher levels found in testis and skeletal muscle (Brown et al. 1994; Chiu et al. 1994).

5 Modular Structure of the TOR Proteins

5.1 FKBP-Rapamycin Binding (FRB) Site

Interestingly, the Tor mutants selected in the original genetic screens for rapamycin resistance were all affected at the same amino acid residues, either Ser1972 in Tor1p or Ser1975 in Tor2p, located just N-terminal to the lipid kinase homology domain (Fig. 2; Cafferkey et al. 1993, 1994; Kunz et al. 1993; Helliwell et al. 1994). Additional, naturally occurring Tor2p mutations conferring partial rapamycin-resistance were later described: Trp2042Leu, Trp2042 Cys, and Phe2049Leu (Lorenz and Heitman 1995). Ser1975, Trp2042, and Phe2049 (in Tor2p) are conserved in Tor1p, Tor2p, and FRAP/mTOR (Lorenz and Heitman 1995). The domain interacting with the FKBP12-rapamycin complex (the FRB domain) was mapped to a region surrounding these residues (amino acids 1886–2081 in Tor2p: Stan et al. 1994; Zheng et al. 1995). Mutating Ser1972 (in Tor1p) or Ser1975 (in Tor2p) to arginine or isoleucine (as was observed in the naturally selected rapamycin-resistant mutants: e.g., Lorenz and Heitman 1995) or to a glutamic acid residue (to mimic serine phosphorylation) prevented interaction with Fpr1p-rapamycin complexes (Zheng et al. 1995). However, phosphorylation of this serine residue is not required for binding to the immunophilin drug complex, as an alanine mutant also supports binding (Zheng et al. 1995). Similar conclusions were reached with the mammalian Tor homologue: mutation of the equivalent serine residue (Ser2035) to isoleucine or threonine, but not to alanine, abolished the interaction with an FKBP-rapamycin complex, and generated dominant rapamycin-resistant mutants. It was thus concluded that the size of the side chain is critical for the interaction of Tor with the FKBP-rapamycin complex: any substitution to an amino acid larger than serine (including the related threonine) abrogated binding. Mutation of other residues (Trp2101 and Phe2108 in the human

FRAP/mTOR) also yielded rapamycin-resistant mutants. The boundaries of the FKBP12-rapamycin binding site in the human protein were further refined, leading to the identification of an 89 amino acid fragment capable of supporting this interaction (Chen et al. 1995). The structure of this domain complexed with FKBP12 and rapamycin was determined by X-ray crystallography, and revealed that rapamycin simultaneously occupies two hydrophobic pockets, one located on FKBP12 and the other located in FRAP/mTOR (Choi et al. 1996). Rapamycin establishes several molecular contacts with both proteins, and is thus able to induce dimerization of FRAP/mTOR and FKBP12 (Choi et al. 1996). Limited contacts occur between the two proteins. The structure also explains the requirement for a small residue at position 2035 (in the human protein): a bulkier residue would prevent the binding of rapamycin in the hydrophobic pocket (Choi et al. 1996).

When overexpressed in *S. cerevisiae*, a small region of Tor1p that binds the Fpr1p-rapamycin complex (but not a mutant unable to bind to the complex) confers rapamycin resistance to the cells, further indicating that the sensitivity of cells to rapamycin is a consequence of the binding to the Fpr1p-rapamycin complex. However, microinjection of the human FRB domain (aa 2015–2114) into osteosarcoma cells prevents G1-S progression (Vilella-Bach et al. 1999). This effect was not observed with a point mutant (Trp2027Phe), indicating that the observed effect is specific. Thus, the mammalian FRB domain appears to act in a dominant-negative fashion, possibly competing with the endogenous FRAP/mTOR for binding to a G1-S effector protein, or a FRAP/mTOR activator (Vilella-Bach et al. 1999).

5.2 Kinase Domain

5.2.1 Location of the Tor Kinase Domains

As mentioned above, the Tor and FRAP/mTOR proteins possess at their C-terminus a portion of significant homology to lipid kinases, especially PI3 kinases (and a weaker homology to PI4 kinases). The FRAP/mTOR kinase homology domain is depicted (domain 4) in Fig. 2. Although initial reports indicated that FRAP/mTOR exhibited PI4 kinase activity, it is now believed that it functions, instead, as a protein kinase. The Tors and FRAP/mTOR are members of a protein family termed the PIKKs (for <u>P</u>hospho<u>I</u>nositide <u>K</u>inase-related <u>K</u>inases), which are large molecular weight proteins sharing a higher level of homology to each other than to PI3 kinases. In all cases, the domain bearing homology to the PI3 kinases is located at the C-terminus of the proteins. Furthermore, all PIKKs share at their extreme C-terminus a short stretch of amino acids (region 5 in Fig. 2), which is not present in PI3 and PI4 kinases.

5.2.2
Phosphoinositide Kinase-Related Kinases (PIKKs)

In mammalian cells, the PIKK family is comprised of FRAP/mTOR, ATM (Ataxia Telangiectasia Mutated), ATR/FRP (Ataxia Telangiectasia and Rad3 related; also known as FRP, FRAP-Related Protein), and DNA-PKc (DNA-activated Protein Kinase, catalytic subunit). Another protein, TRRAP (Transformation/Transcription domain-Associated Protein) also exhibits significant homology to the PIKKs, although it lacks conserved residues in the catalytic domain (and is therefore not expected to possess kinase activity). Most PIKKs are conserved in other species: the *S. pombe* Rad3 is most closely related to ATR/FRP, the *S. cerevisiae* Esr1p/Mec1p, the *Aspergillus nidulans* UVSB PI3 kinase and to the *Drosophila melanogaster* Mei-41 protein (especially in the C-terminal portion of the proteins). Similarly, the *S. pombe* Tel1 bears strong homology to ATM and to an *S. cerevisiae* Tel1p homologue, again, with the strongest homology in the C-terminus (although functionally, ATM appears closer to Mec1p and Rad3 than to Tel1p). TRRAP also possesses homologues in several species, including *S. cerevisiae* (Tra1p), *S. pombe* and *D. melanogaster*. A notable exception may be the DNA-PKc protein, which appears to be restricted to vertebrates. The other PIKK family members have not been implicated in translational control, but are involved in checkpoint control following DNA damage. The reader is directed to several recent reviews on this topic (Lavin and Shiloh 1997; Shiloh 1997; Canman and Lim 1998; Critchlow and Jackson 1998; Jeggo et al. 1998; Rotman and Shiloh 1998, 1999; Featherstone and Jackson 1999; Smith et al. 1999a, b).

5.2.3
Role of the Kinase Domain in Tor Function

As mentioned above, FRAP/mTOR, despite its homology to PI3 kinases, functions as a protein kinase. Immunoprecipitates of endogenous FRAP/mTOR, or FRAP/mTOR expressed in insect cells, autophosphorylate on Ser2481 (Brown et al. 1995; Brunn et al. 1996; Peterson et al. 2000). FRAP/mTOR immunoprecipitates have also been reported to phosphorylate other protein targets, including p70S6 kinase and the translation inhibitors 4E-BP1 and 4E-BP2 (reviewed in Raught et al. 2000a). An intact kinase domain is necessary for FRAP/mTOR protein function. Mutations at positions corresponding to those that abrogate the lipid kinase activity of PI3 K prevent FRAP/mTOR autophosphorylation, and its ability to signal to p70S6 kinase and 4E-BP1. While a mutation at Ser2035 (in FRAP/mTOR) confers rapamycin resistance, abrogation of kinase activity by addition of a second mutation in the kinase domain prevents Ser2035 mutants from acting as dominant rapamycin-resistant mutants (e.g., Brown et al. 1995; Brunn et al. 1997b). The kinase activity of the *S. cerevisiae* Tor proteins is also essential to their function. Overexpression in a wild-type background of a kinase-dead Tor1p confers a dominant-negative phenotype, leading to G1 arrest (Zheng et al. 1995). The kinase domain of Tor2p is also required for the function of the protein, since a protein possessing a mutation in the catalytic domain is unable to rescue yeast deleted for *TOR2* (Zheng et al. 1995). Taken together, these data indicate that an intact kinase domain is essential for the function of the Tor proteins. However, recent evidence suggests that the predicted kinase domain of FRAP/mTOR alone is insufficient for autokinase activity, as discussed below (Sect. 5.3). A fragment located N-terminal to the kinase domain (containing the FKBP-binding site) is also necessary for kinase activity (Vilella-Bach et al. 1999).

5.2.4
Inhibition of Tor Protein Kinase Activity

Wortmannin and LY294002 are two structurally unrelated molecules which, at low concentrations, are relatively specific PI3 kinase inhibitors (Powis et al. 1994; Vlahos et al. 1994; Ui et al. 1995). Wortmannin is an irreversible PI3 kinase inhibitor, which functions by covalently binding to and modifying a lysine residue in the catalytic site of PI3 kinase (Wymann et al. 1996). LY294002 is a reversible inhibitor of PI3 kinase, which has been increasingly used due to its superior stability in biological systems, as compared to wortmannin (Vlahos et al. 1994). Because of its similarity to the kinase domain of PI3Ks, the direct inhibition of FRAP/mTOR kinase activity by wortmannin and LY294002 was also investigated (Brunn et al. 1996). Radiolabeled wortmannin was demonstrated to bind to FRAP/mTOR, and this binding was prevented by pre-incubation with a non-hydrolyzable ATP analog (Brunn et al. 1996). Consistent with this, wortmannin inhibits FRAP/mTOR autokinase activity in vitro, with an IC_{50} of approximately 200 nM, which is about 100-fold more than the amount required for PI3 kinase inhibition (Brunn et al. 1996). In Sf9 insect cells expressing recombinant FRAP/mTOR, treatment with wortmannin prior to cell lysis and kinase assay also inhibited FRAP/mTOR kinase activity (with an IC_{50} of 300 nM), indicating that FRAP/mTOR can be inhibited by wortmannin in vivo (Brunn et al. 1996). LY294002 was also demonstrated to inhibit FRAP/mTOR autokinase activity in vitro, with an IC_{50} of 5 μM (Brunn et al. 1996). The other members of the PIKK family are also sensitive to wortmannin treatment: wortmannin inhibits DNA-PKc with a published IC_{50} of approximately 250 nM (Hartley et al. 1995) or 16 nM (Sarkaria et al. 1998). ATM is inhibited by wortmannin with an IC_{50} of about 150 nM, but ATR is less sensitive to wortmannin inhibition, with an IC_{50} of approximately 1.8 μM (Sarkaria et al. 1998). Consistent with the concentrations required to inhibit ATM and DNA-PKc, wortmannin treatment also leads to radiosensitization (Sarkaria et al. 1998). Thus, inhibition of the PIKKs by wortmannin and LY294002 should be considered when analyzing the effects of these inhibitors: in some cases, events which have been attributed to PI3 kinase inhibition may actually be due to inhibition of one or more PIKKs, including FRAP/mTOR.

While rapamycin complexed with FKBP12 potently inhibits downstream signaling by FRAP/mTOR in vivo, the issue of whether rapamycin affects the kinase activity of FRAP/mTOR itself is controversial. Addition of rapamycin and FKBP12 to a FRAP/mTOR immunoprecipitate inhibits FRAP autokinase activity in vitro, but the concentrations required for this effect are very high as compared to the concentrations of rapamycin required to inhibit FRAP/mTOR signaling in vivo. Furthermore, addition of rapamycin to cells does not affect the autophosphorylation of FRAP/mTOR in vivo at Ser2481, as determined by the use of a phosphospecific antibody directed against this site (Peterson et al. 2000). In contrast, wortmannin treatment completely prevents this phosphorylation event (Peterson et al. 2000). These data are consistent with results from Tor2p kinase mutants in *S. cerevisiae*: a Tor2p kinase-dead mutant is unable to rescue the lethality associated with *TOR2* disruption. However, rapamycin treatment of yeast is not lethal, but instead arrests the cells in G1 (as described above). If rapamycin was acting on Tor2p to inhibit its kinase activity, it would be expected that the phenotypes of rapamycin addition and of the mutation in the Tor2p kinase domain would be equivalent. Taken together, these data argue against a direct inhibition of the kinase activity of the Tor proteins by rapamycin.

5.3
Other Structural Elements in TOR and FRAP/mTOR

The Tor proteins also possess two other, less well characterized structural elements. The first 1200 amino acids of the Tor and FRAP/mTOR proteins comprise a "HEAT" motif, named for Huntingtin, Elongation factor 3, the regulatory A subunit of PP2 A and Tor1p, the first proteins found to possess this domain (Andrade and Bork 1995; Groves et al. 1999a, b). This motif consists of stretches of about 40 amino acids in at least three repeats, and displays a consensus pattern of hydrophobic, proline, aspartic acid, and arginine residues, suggesting a common architecture. HEAT repeats have also been detected in several other proteins, many of which function as adapters or scaffolds. The crystal structure of the adapter (A) subunit of PP2A, which is composed exclusively of these repeats, resembles a curved rod, and consists entirely of repeated alpha helix-loop-alpha helix motifs (Groves et al. 1999b). The HEAT domain is presumed to be involved in mediating protein-protein interactions. For example, the adapter (A) subunit of PP2A is responsible for binding to both the catalytic (C) and the regulatory (B) PP2A subunits, and also binds to the T antigens of several different viruses. Similarly, the HEAT motifs of importin ß create a docking site for the binding of importin α and Ran-GTP. A protein interacting with a portion of the FRAP/mTOR HEAT domain (aa 1017–1046) has recently been isolated by two-hybrid screening (Sabatini et al. 1999). This molecule, termed gephyrin, is a tubulin-binding protein involved in postsynaptic clustering of neuronal glycine receptors (Sabatini et al. 1999). Gephyrin is presumed to modulate the intracellular localization of FRAP/mTOR, and mutations that abrogate gephyrin binding prevent FRAP/mTOR signaling to two downstream translational targets, the ribosomal S6 kinase 1 (S6K1) and eIF4E-binding protein 1 (4E-BP1; see below).

Another conserved region in the Tor proteins, initially termed the central toxic effector domain, spans roughly 750 aa, and inhibits yeast growth when overexpressed (Fig. 2). This portion of Tor is located immediately C-terminal to the HEAT domain and N-terminal to the FRB domain. How this portion of Tor confers a toxic effect in yeast is unknown, although it is presumed to interfere with the activity of the endogenous FRAP/mTOR by titrating out an effector or activator (Alarcon et al. 1999). This "toxic" effect may not be restricted to the Tor or FRAP/mTOR proteins. The same region of the protein (the boundaries being amino acids 1382–1982 in the human FRAP/mTOR) was also described as a novel PIK-related kinase domain, termed FAT, for FRAP, ATM and TRAPP (Alarcon et al. 1999; Bosotti et al. 2000). The PIKK kinases and the TRAPP proteins share roughly 16% average identity within this region (as opposed to 28% in the kinase domain). Since this domain is always found in combination with the C-terminal PIKK-specific region, intramolecular interactions between the FAT domain and the C-terminal region were postulated to be involved in modulating kinase activity. Interestingly, deletion analysis of the N-terminus of FRAP/mTOR revealed that the kinase homology domain alone did not exhibit autokinase activity in vitro (or kinase activity directed toward 4E-BP1). Inclusion of N-terminal sequences containing the FRB domain was also not sufficient to restore kinase activity (Vilella-Bach et al. 1999). Only when the FAT domain was included was the autokinase activity and the kinase activity toward 4E-BP1 restored (Vilella-Bach et al. 1999).

6 Signaling to TOR (FRAP/mTOR): Activation by Extracellular Stimuli

Study of the regulation of FRAP/mTOR by upstream stimuli has been slowed by the absence of a system allowing for detection of changes in its activity. FRAP/mTOR activity is usually monitored by immunoprecipitating the kinase from cells treated with various stimuli or inhibitors, then measuring autokinase activity or kinase activity toward exogenous substrates (most often 4E-BP1 and S6K1; see below). An obvious caveat to these experiments is that only FRAP/mTOR kinase activity directed toward selected substrates is detected by these methods, while the biological activities of FRAP/mTOR may not be limited to these events in vivo (and for that matter may not even be limited to the protein kinase activity of FRAP/mTOR). Thus, results describing "activation" or "inactivation" of FRAP/mTOR by stimuli or inhibitors must be interpreted with caution.

In this regard, two hypotheses (which are not mutually exclusive) have been proposed regarding the regulation of FRAP/mTOR activity: (1) FRAP/mTOR is activated directly by growth factors (or other stimuli), then signals to various downstream targets, or (2) FRAP/mTOR acts as a gatekeeper to "sense", for example, amino acid sufficiency (see below). In case 2, FRAP/mTOR has been predicted to remain constitutively active in the presence of sufficient nutrients, a lack of nutrients is predicted to inhibit FRAP/mTOR signaling.

Modest changes in FRAP/mTOR kinase activity (~1.5–3-fold increase in activity toward 4E-BP1) have been reported following treatment of cells with serum or mitogens (Scott et al. 1998; Scott and Lawrence 1998; Navé et al. 1999; Sekulic et al. 2000), or following activation of a kinase that is postulated to lie upstream of FRAP/mTOR, Akt/PKB. Activation of the kinase activity directed toward exogenous substrates parallels a modulation of ^{32}P incorporation into the FRAP/mTOR protein itself (Scott et al. 1998). As previously mentioned, FRAP/mTOR autophosphorylates on a serine residue in vitro, and is phosphorylated on serine in vivo (e.g., Peterson et al. 2000). More recently, the study of FRAP/mTOR phosphorylation in vivo using phosphospecific antibodies has indicated that the autophosphorylation site is located at Ser2481 (Peterson et al. 2000), and an Akt/PKB phosphorylation site at Ser2448 (Navé et al. 1999; Sekulic et al. 2000). Since phosphorylation of at least one FRAP/mTOR target, 4E-BP1, also occurs downstream of Akt/PKB (Gingras et al. 1998; Kohn et al. 1998; Dufner et al. 1999; Takata et al. 1999), this kinase was postulated to signal upstream from FRAP/mTOR. In a cell line expressing a conditionally activated Akt/PKB protein, Akt/PKB activation elicits an increase in the immunoprecipitable kinase activity associated with FRAP/mTOR (Scott et al. 1998). Akt/PKB can phosphorylate FRAP/mTOR directly on Ser2448 in vitro, and this phosphorylation event is responsive to insulin and wortmannin treatments in vivo (Navé et al. 1999; Sekulic et al. 2000). However, the physiological role of Ser2448 phosphorylation is unclear, because phosphorylation of this residue is not necessary for signaling to either 4E-BP1 or S6K1 (Sekulic et al. 2000). It has also been recently reported that PKCδ associates with FRAP/mTOR and signals to 4E-BP1 phosphorylation (Kumar et al. 2000a). Whether PKCδ activates FRAP/mTOR is unknown. A role of the kinase c-Abl in signaling to FRAP/mTOR has also been proposed: phosphorylation of FRAP/mTOR by c-Abl was postulated to negatively regulate it, resulting in a dephosphorylation of downstream FRAP/mTOR targets (Kumar et al. 2000b).

7
Downstream of Tor: Signaling Through Phosphatases

Signaling cascades are often represented by a series of unidirectional arrows, indicating the successive action of protein kinases. Less studied and seldom depicted in these models, the action of phosphatases (especially serine/threonine phosphatases) appears underestimated. Phosphatases are also crucial components of signaling cascades, and, like kinases, their action leads to activation or inactivation of downstream targets. Protein serine/threonine phosphatases may play an important role in rapamycin-sensitive signaling. Genetic screening in *S. cerevisiae* has identified several phosphatases or phosphatase-associated proteins as part of the rapamycin-sensitive signaling pathway. Deletion or mutation of the PP2A regulatory subunits *CDC55* and *TPD3* (Jiang and Broach 1999) or overexpression of the PP2A-related phosphatase Sit4p (Di Como and Arndt 1996) confers partial resistance to rapamycin in yeast. Mutations in a phosphatase-associated protein termed Tap42p also confer rapamycin-resistance (Di Como and Arndt 1996). *S. cerevisiae* expressing temperature-sensitive mutants of the Tap42p protein also exhibit a dramatic defect in translation initiation when grown at the non-permissive temperature (Di Como and Arndt 1996). Tap42p interacts with the catalytic subunit of PP2A, as well as with other PP2A-related phosphatases such as Sit4p (Di Como and Arndt 1996). While PP2A-type phosphatases normally function as trimeric heterocomplexes, in which the catalytic (c) subunit is bridged to the regulatory (b) subunit through an adapter (a) protein, the interaction of Tap42p with PP2A involves direct binding to the catalytic (c) subunit in a heterodimeric complex (Di Como and Arndt 1996; Jiang and Broach 1999). The PP2Aa-PP2Ab complex competes with Tap42p for binding to PP2Ac, and their binding to PP2Ac is mutually exclusive. The association of Tap42p with PP2Ac is disrupted by nutrient deprivation or rapamycin treatment, and has been linked to Tor2p activity because rapamycin-sensitivity can be relieved by expression of a rapamycin-resistant Tor2p mutant protein (Di Como and Arndt 1996). The role of Tor1/2p in regulating the association between Tap42p and PP2Ac appears to be through modulation of Tap42p phosphorylation. Tap42p is a phosphoprotein, and Tap42p phosphorylation is sensitive to rapamycin (Jiang and Broach 1999). Tap42p rapamycin-sensitivity is abrogated in a strain expressing a rapamycin-resistant Tor1p protein (Jiang and Broach 1999). Furthermore, in vitro, a Tor1p immunoprecipitate can phosphorylate Tap42p (Jiang and Broach 1999), suggesting that Tor1p (or an associated kinase) directly regulates its binding to PP2A-type phosphatases. Dephosphorylation of Tap42p appears to be mediated by the trimeric PP2A complex, as mutations in the PP2Aa (*TPD3*) or PP2Ab (*CDC55*) subunits prevent dephosphorylation of Tap42p following rapamycin treatment (Jiang and Broach 1999). Together, these observations have led to the model that Tap42p phosphorylation is positively regulated by the Tor proteins, and is necessary for its interaction with PP2Ac, and that the Tap42p-PP2Ac complex is a positive effector in Tor signaling to translation initiation.

Homologues of Sit4p (the phosphatase PP6), PP2A and Tap42p (the B cell receptor binding protein, α4) have been identified in mammalian cells. The interaction between α4 and phosphatases is also conserved in mammals; α4 binds directly to the catalytic subunits of PP2A (Inui et al. 1998; Murata et al. 1997), PP4 and PP6 (Chen et al. 1998; Nanahoshi et al. 1999). However, how α4 or Tap42p modulate their binding partners is not well understood; binding has been reported to both increase and decrease phosphatase activity, or to alter substrate specificity (Murata et al. 1997; Nanahoshi et al. 1998). Like Tap42, α4 was demonstrated to be a phosphoprotein in vivo, and the α4-PP2A interaction was reported to be significantly inhibited by rapamycin (Murata et al. 1997; Inui et al. 1998).

8 Tor Proteins as Sensors of Nutrient Availability

Rapamycin treatment, or combined deletion of the *TOR1* and *TOR2* genes in *S. cerevisiae* yields a phenotype similar to that observed upon starvation: severe reduction in protein synthesis (see Sect. 10), vacuolar enlargement, accumulation of glycogen, acquisition of thermotolerance, induction of autophagy (see below), and cell cycle arrest early in G1 (G0) (Heitman et al. 1991a; Kunz et al. 1993; Cardenas and Heitman 1995; Barbet et al. 1996; Noda and Oshumi 1998). Because of the similarity between the phenotype of rapamycin-treated yeast cells and starved yeast, the involvement of the Tor proteins in nutrient-induced signaling has been extensively studied. In recent years, use of DNA arrays has linked transcription of several nutrient-sensitive genes to Tor signaling (Cardenas et al. 1999; Hardwick et al. 1999). This transcriptional upregulation of the nutrient-responsive genes involves Tor-dependent phosphorylation of the transcription factor Gln3p and of its repressor Ure2p (Beck and Hall 1999a; Cardenas et al. 1999; Hardwick et al. 1999). A link between the Tor proteins and amino acid transport has also been established (Schmidt et al. 1998; Beck et al. 1999b). The effects on amino acid transport and nutrient-sensitive transcription are beyond the scope of this review, and the reader is referred to the original publications.

9 FRAP/mTOR and TOR as Translational Regulators

9.1 FRAP/mTOR Effects on Translation in Mammalian Cells

As mentioned earlier, in mammalian cells, FRAP/mTOR modulates the phosphorylation state of several translation initiation factors, inhibitors, and other proteins involved in translation. Rapamycin prevents S6K1 activation (reviewed in Jefferies et al. 1994; Fumagalli and Thomas 2000). Rapamycin also affects phosphorylation of one subunit of eIF4F, eIF4GI, of the initiation factor eIF4B and of the translation inhibitors 4E-BPs (Fig. 3; Graves et al. 1995; Lin and Lawrence 1996; Raught et al. 2000a, b). The role of these translation factors and inhibitors, and the signaling from FRAP/mTOR to 4E-BP1 are addressed in Section 10.

In mammalian cells, the effects of rapamycin on translation differ according to cell line. Cap-dependent translation in NIH 3T3 cells is decreased by about 40% (after a 20-h treatment), while cap-independent translation is unaf-

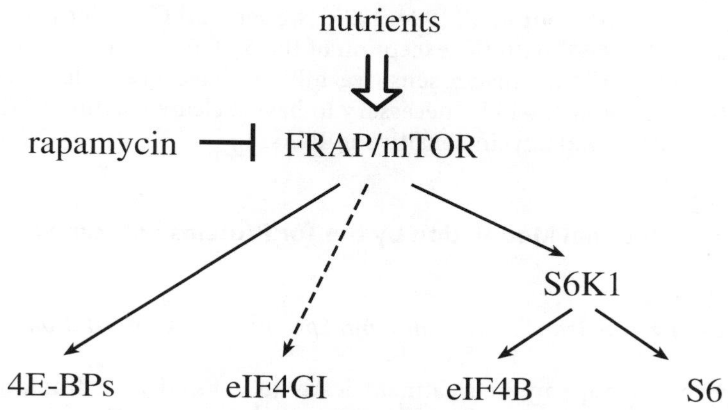

Fig. 3. Nutrients signal to translation initiation through FRAP/mTOR. In the presence of amino acids, FRAP/mTOR signals to translation initiation by altering the phosphorylation of several components of the translational machinery. The changes in phosphorylation positively correlate with the translation rates. In the case of the translational inhibitors 4E-BP1 and 4E-BP2, hyper-phosphorylation results in a reduced binding to eIF4E and a relief of translational repression. For S6K1, nutrient signaling induces an increase in kinase activity. S6K1 phosphorylates at least two components of the translational machinery, the ribosomal S6 protein and eIF4B, but the molecular function of the phosphorylation of these two substrates is unknown. eIF4GI is also phosphorylated on three residues in a rapamycin-dependent manner, but the effect of this phosphorylation event and the nature of the kinase directly responsible remain to be elucidated. *Filled lines* indicate that a direct phosphorylation in vitro has been reported (in the case of FRAP/mTOR, this was performed using immunoprecipitated kinase). *Dashed lines* represent a lack of in vitro data

fected by this treatment (e.g., Beretta et al. 1996). Overall translation rates in other cell lines are only mildly affected by rapamycin (e.g., Jefferies et al. 1994). However, the translation of certain specific mRNAs is extremely sensitive to rapamycin (e.g., Pedersen et al. 1997). The best studied example is a class of mRNAs possessing a 5′ terminal oligopyrimidine tract (5′TOP), whose translation is nearly abolished by rapamycin. Most 5′TOP-containing mRNAs are components of the translational machinery: ribosomal proteins and elongation factors (reviewed in Jefferies and Thomas 1996; Meyuhas et al. 1996; Meyuhas and Hornstein 2000). Translation of these mRNAs is modulated through S6K1, which phosphorylates the ribosomal protein S6 and probably at least one translation initiation factor, eIF4B (reviewed in Fumagalli and Thomas 2000; Raught et al. 2000a). Other rapamycin-sensitive mRNAs include the ornithine decarboxylase (ODC) and c-myc mRNAs, a group characterized by long, structured 5′ UTRs. Such mRNAs are normally translated poorly in cells, presumably because of the limiting amount of eIF4F, the helicase activity of which is required for melting inhibitory secondary structure. A third example of rapamycin-sensitive mRNA is the IGFII leader 3 translation, which appears to be rapamycin-sensitive due to the binding of specific proteins to its

5′ UTR (De Moor et al. 1994, 1995; Nielsen and Christiansen 1995; Nielsen et al. 1995, 1999). With the exception of the 5′TOP-containing mRNAs, only a few examples of rapamycin-sensitive mRNAs have been identified. A genomic-based approach will be necessary to have a clearer picture of the number and identity of rapamycin-sensitive mRNAs.

9.2 Translational Modulation by the Tor Proteins in *S. cerevisiae*

9.2.1 Changes in Translation Rates and Specific mRNA Translation

In yeast, rapamycin treatment leads to a rapid and dramatic translational arrest, accompanied by polysomal disaggregation (Barbet et al. 1996). The translational arrest occurs prior to, and is presumed to be the cause of, cell cycle arrest in G1 (Barbet et al. 1996). Cycle arrest in rapamycin-treated cells could be due to a decrease in cyclin 3 (*CLN3*) mRNA translation. Cln3p activates the transcription of late G1-specific genes including *CLN1* and *CLN2*, and is translationally regulated through the presence of an upstream open-reading-frame (uORF) in its 5′ UTR (Polymenis and Schmidt 1997). uORFs are normally inhibitory to translation (reviewed in Geballe and Sachs 2000). *CLN3* mRNA translation is rapidly repressed following rapamycin treatment or nitrogen deprivation (Gallego et al. 1997). The cycle arrest (but not the general decrease in translation initiation) can be overcome if the *CLN3* coding sequence is fused to the promoter and the 5′ UTR of the polyubiquitin-encoding gene *UBI4*. The presence of the UBI4 5′ UTR renders *CLN3* translation less dependent upon eukaryotic initiation factor eIF4E (*CDC33*). Similarly, mutation of the initiator codon (AUG) of the uORF alleviates the translational block of *CLN3* and restores cell cycle progression. In *S. cerevisiae* mutants defective in eIF4E (which also arrest in G1), expression of *CLN3* is sufficient to restore the G1 to S phase transition (Danaie et al. 1999).

9.2.2 Putative Effectors of the Translational Effect of the Tor Proteins

The translational effects of Tor proteins appear to be mediated, at least in part, through the Tap42p protein, as polysomes are strongly reduced in a *tap42* mutant grown at the non-permissive temperature (Di Como and Arndt 1996). The exact nature of the translational target (or targets) of rapamycin in yeast is not known, but a few candidates have been put forward. Rapamycin induces a rapid and specific degradation of the yeast translation factors eIF4G1 and eIF4G2 (Tif4631p and Tif4632p), which is prevented upon transformation with a dominant rapamycin-resistant mutant of *tor1* (Berset et al. 1998; Powers and Walter 1999). The rate of degradation of eIF4G proteins following rapamycin treatment was paralleled by translational inhibition (Powers and Walter 1999).

Whether eIF4G degradation following rapamycin treatment is a general effect or a strain-specific effect is not clear, however. Another putative candidate for mediating at least some of the rapamycin effects is Eap1p, which was cloned based on its ability to interact with the yeast eIF4E protein (Cosentino et al. 2000). While no change in affinity for eIF4E was reported for Eap1p in the presence of rapamycin, deletion of *EAP1* conferred a mild rapamycin resistance phenotype (Cosentino et al. 2000).

9.2.3
Changes in the Biosynthesis of the Translational Apparatus

Effects of rapamycin on ribosome biosynthesis in yeast have also been reported. The Tor pathway was implicated in the regulation of the activity of RNA Pol I and Pol III enzymes (Zaragoza et al. 1998), consistent with the inhibition of rRNA synthesis and processing observed following addition of rapamycin (Powers and Walter 1999). Furthermore, rapamycin treatment decreases ribosomal protein mRNA transcription early after addition of the drug (Cardenas et al. 1999; Hardwick et al. 1999; Powers and Walter 1999). The mRNA levels for several translation factors tested also decrease, albeit more slowly, following addition of rapamycin to yeast (Powers and Walter 1999). Taken together, these data clearly indicate that following rapamycin treatment the rate of synthesis of components of the translational machinery is reduced.

10
Regulation of 4E-BP1 Activity by FRAP/mTOR

10.1
Involvement of 4E-BP1 in Translation Initiation

Translation initiation in eukaryotes is facilitated by the structure m^7GpppX (where X is any nucleotide) located at the 5′ end of most cellular mRNAs. This structure, commonly known as the mRNA "cap", is specifically recognized by the eukaryotic translation initiation factor 4F (eIF4F). eIF4F is a trimeric complex composed of a cap binding protein (eIF4E), an RNA helicase (eIF4 A), and a scaffolding protein (eIF4G). The activity and regulation of the eIF4F complex have been reviewed in detail elsewhere (Gingras et al. 1999a; Hershey and Merrick 2000; Raught et al. 2000a). In recent years, a small family of proteins that inhibit translation by binding to eIF4E and preventing its incorporation into eIF4F has been intensively studied. The eIF4E binding proteins, 4E-BP1, 4E-BP2, and 4E-BP3 (eIF4E-Binding Proteins 1, 2 and 3), share more than 55% identity at the amino acid level, and inhibit cap-dependent, but not cap-independent, translation both in vitro and in vivo (Pause et al. 1994; Poulin et al. 1998). Binding of the 4E-BPs to eIF4E prevents the association between eIF4E and eIF4G and, thus, the assembly of a functional eIF4F complex (reviewed in Gingras et al. 1999a; Raught et al. 2000a). The interaction with eIF4E is conferred by a highly conserved amino acid motif present both in the 4E-BPs and eIF4Gs, which interacts with the convex dorsal surface of eIF4E. As determined by X-ray crystallography, the 4E-BPs act as molecular mimics of the eIF4E binding site in the eIF4G proteins (Marcotrigiano et al. 1999).

10.2 4E-BP1 Phosphorylation Sites

Phosphorylation of specific serine and threonine residues modulates the affinity of the 4E-BPs for eIF4E (e.g., Lin et al. 1994; Pause et al. 1994; Fadden et al. 1997). While hypophosphorylated 4E-BPs bind efficiently to eIF4E, phosphorylation of a critical set of residues abrogates this interaction (e.g., Lin et al. 1994; Pause et al. 1994; Fadden et al. 1997). 4E-BP phosphorylation levels are increased by many types of extracellular stimuli, including hormones, growth factors, cytokines, mitogens, and G-protein coupled receptor ligands (see Table 2 in Gingras et al. 1999a). It is noteworthy that 4E-BP1 was first described in the early 1980s as a protein which is highly phosphorylated after insulin stimulation of rat adipocytes (Belsham and Denton 1980; Belsham et al. 1982; Blackshear et al. 1982, 1983). This protein was later biochemically purified and cloned, and termed PHAS-I, for Phosphorylated Heat and Acid Stable protein-Insulin responsive (Hu et al. 1994). The function of PHAS-I was revealed when it was found to be the rat ortholog of the human 4E-BP1. Six phosphorylation sites have been identified in the mammalian 4E-BP1 protein (Fig. 2; Fadden et al. 1997; Heesom et al. 1998). Five of the six sites are followed by a proline residue, and one is followed by a glutamine. Two phosphorylated residues, Thr37 and Thr46, lie on the amino terminal side of the eIF4E binding motif (located at aa 54–60), and four phosphorylated residues have been identified on the carboxy terminal side of the eIF4E binding motif: Ser65, Thr70, Ser83 and Ser112. The 4E-BP1 residues phosphorylated in vivo may vary somewhat according to cell type and/or species.

10.3 FRAP/mTOR Signals Upstream of 4E-BP1

The identification of 4E-BP1 as a target of FRAP/mTOR stems from the initial observation that rapamycin decreases the phosphorylation state of 4E-BP1 and increases its affinity for eIF4E. Interestingly, the extent of inhibition of phosphorylation by rapamycin on the various 4E-BP1 phosphorylation sites varies. In 293 cells, the phosphorylation of Thr37 and Thr46 is relatively resistant to inhibition by rapamycin, whereas the phosphorylation of Ser65 and Thr70 is nearly abolished following rapamycin treatment (Gingras et al. 1999b, 2001). The study of rat 4E-BP1 Ser111 (corresponding to human Ser112) led to the conclusion that this site was insensitive to rapamycin (but is sensitive to the PI3 kinase inhibitor wortmannin; Heesom et al. 1998). With the exception of Ser112, the sensitivity of the various sites to rapamycin reflects the sensitivity of these sites to serum starvation, or treatment with inhibitors of PI3 kinase. The signaling pathway to 4E-BP1 has been described elsewhere (Gingras et al. 1999a; Raught et al. 2000a), and involves, under most circumstances, activation of PI3 kinase and a downstream effector, the proto-oncogene Akt/PKB. Akt/PKB was shown to induce 4E-BP1 phosphorylation on the same sites exhibiting sensitivity to serum-treatment, Ser65 and Thr70 (Gingras et al. 1998). PI3 kinase and Akt/PKB have little effect (at least in 293 cells) on the phosphorylation of Thr37 and Thr46.

4E-BP1 phosphorylation is an ordered, hierarchical process. Mutation of Thr37 or Thr46 (or both residues) to alanine(s) abolishes phosphorylation of Ser65 and Thr70 in serum-replete 293 T cells (Gingras et al. 1999b). Phosphorylation on Ser65 and Thr70 in endogenous 4E-BP1 is only detected in protein species also phosphorylated on Thr37 and Thr46 (Gingras et al. 2001). Thus, phosphorylation at Thr37 and Thr46 appears to act as a "priming" event for the phosphorylation of Ser65 and Thr70. How phosphorylation at Thr37 and Thr46 may act as a priming event has not been elucidated; it is possible that it directly recruits a Thr70/Ser65 kinase, or it could induce a conformational change at the eIF4E/4E-BP1 interface to facilitate access to a kinase.

Expression of a rapamycin-resistant FRAP/mTOR mutant protein confers rapamycin resistance to 4E-BP1 phosphorylation (Brunn et al. 1997b; Hara et al. 1997; Gingras et al. 1998). Immunoprecipitates of FRAP/mTOR phosphorylate 4E-BP1 on the "priming" sites, Thr37 and Thr46 in 4E-BP1 (Brunn et al. 1997a, b; Burnett et al. 1998; Gingras et al. 1999b; Heesom and Denton 1999). However, it is likely that FRAP/mTOR also plays a critical regulatory role in the phosphorylation of Ser65 and Thr70, since these residues display a higher level of rapamycin sensitivity than Thr37 and Thr46. Presumably, FRAP/mTOR modulates the activity of a Ser65/Thr70 kinase or phosphatase (Gingras et al. 1999b, 2001). PP2A (or PP2A-like phosphatases) could be involved in 4E-BP1 dephosphorylation induced by rapamycin, because treatment of cells with the phosphatase inhibitor calyculin prevents 4E-BP1 dephosphorylation (Peterson et al. 1999). It is thus possible that a phosphatase activity directed toward Ser65 and Thr70 (the most rapamycin-sensitive sites) is activated following rapamycin treatment. It has been reported that in vitro both the Tap42 protein and α4 interfere with PP2A-induced 4E-BP1 dephosphorylation in vitro (Nanahoshi et al. 1998), but the physiological consequences of this inhibition have not been addressed.

11
A Role for FRAP/mTOR in Signaling to 4E-BP1 in Response to Nutrient Availability

A striking reduction in translation initiation rates occurs in cultured mammalian cells deprived of amino acids. Re-addition of amino acids to the culture media readily reverses the translation inhibition, indicating that the nutrients themselves play a role in translational control. Numerous reports have indicated that, in mammalian cells, amino acid deprivation modulates the activity of several proteins involved in translation, including 4E-BP1 (reviewed in Kimball and Jefferson 2000). Incubation of cells in culture medium lacking amino acids results in a reduction of the basal phosphorylation state of 4E-BP1 (Hara et al. 1998). Re-addition of amino acids alone can partially restore 4E-BP1 phosphorylation, and amino acids synergize with insulin, IGF I or serum to elicit complete phosphorylation (Hara et al. 1998; Xu et al. 1998a, b).

Further studies have implicated specific amino acids in the stimulation of 4E-BP1 phosphorylation: leucine (and, to a lesser extent, other branched-chain amino acids) is necessary for 4E-BP1 phosphorylation. Leucine alone can also stimulate, at least partially, the phosphorylation of 4E-BP1 in amino acid-deprived cells. The specificity of leucine in mediating the increase in 4E-BP1 phosphorylation suggests the existence of a specific, and still unknown "receptor" for leucine. While PI3 kinase and Akt/PKB are activated and essential for 4E-BP1 phosphorylation following serum or insulin treatment, amino acid removal or re-addition did not alter the activity of PI3 kinase or Akt/PKB, or the activation of these proteins by serum or insulin (Hara et al. 1998; Patti et al. 1998; Campbell et al. 1999; Kimball et al. 1999; Shigemitsu et al. 1999). These results indicate that the signaling pathway to 4E-BP1 phosphorylation modulated by amino acids does not involve PI3 kinase or Akt. While amino acid-induced 4E-BP1 phosphorylation is sensitive to wortmannin treatment (Patti et al. 1998; Wang et al. 1998; Xu et al. 1998a), higher concentrations of wortmannin are necessary for inhibiting S6K1 and 4E-BP1 phosphorylation after amino acid stimulation, as compared to the amount necessary to inhibit phosphorylation of these proteins after insulin stimulation (Shigemitsu et al. 1999). These data are consistent with the hypothesis that the target of wortmannin following amino acid-stimulation is not PI3 kinase, but another kinase with a lower sensitivity to wortmannin. Since the PIKKs (including FRAP/mTOR) are inhibited by wortmannin at higher concentrations than those required for PI3 kinase inhibition (Brunn et al. 1996; Sarkaria et al. 1998), it is possible that the inhibition observed with wortmannin after amino acid stimulation is due to inhibition of FRAP/mTOR itself. The involvement of FRAP/mTOR in amino acid signaling was also suggested by the observation that a rapamycin-resistant S6K1 protein is also insensitive to amino acid withdrawal (Hara et al. 1998), and a rapamycin-resistant FRAP/mTOR protein confers resistance to amino acid deprivation in the wild-type S6K1 (Iiboshi et al. 1999). Thus, FRAP/mTOR may play an important checkpoint role, allowing propagation of intracellular signals to the translational apparatus only when sufficient amino acids are present.

12 FRAP/mTOR as a Mediator of "Translational Homeostasis"

Treatment of cells with translational inhibitors such as anisomycin or cycloheximide leads to a compensatory hyperphosphorylation of both 4E-BP1 and S6K1 (e.g., Brown and Schreiber 1996; von Manteuffel et al. 1996). By contrast, in a murine fibroblast cell line transformed by eIF4E overexpression (Lazaris-Karatzas et al. 1990), in which general translation rates are increased (Rosenwald et al. 1999), both 4E-BP1 and S6K1 are maintained in a hypophosphorylated state, as compared to untransformed cells (Khaleghpour et al. 1999). To determine whether the basis for this phenomenon is eIF4E overexpression itself, and not the transformed state it induces, tetracycline-inducible eIF4E overexpressing fibroblast lines were generated (Khaleghpour et al. 1999). Upon removal of tetracycline from the cell culture media, these cell lines overexpress eIF4E to varying levels. Strikingly, the degree of eIF4E overexpression was observed to correlate with the degree of 4E-BP1 and S6K1 dephosphorylation (Khaleghpour et al. 1999). 4E-BP1 was also observed to be hypophosphorylated in several murine mammary tumor cell lines, in contrast to non-tumorigenic parental cell strains (Raught et al. 1996). These data are consistent with a model in which an unscheduled change in translation initiation rates is sensed by the cell, and the signaling pathways modulating 4E-BP1 and S6K1 activity respond in a compensatory manner. It is also tempting to speculate that transformed cells may acquire the ability to bypass this translational inhibition mechanism.

13 Future Prospects

Over the past 10 years, much has been elucidated regarding the pathway inhibited by rapamycin in yeast and mammalian cells. Tor proteins have been linked to the propagation of signals from nutrient availability to translation, amino acid import, specific mRNA transcription, and ribosome biosynthesis. Extensive genetic analysis in yeast has allowed the identification of several downstream Tor targets, most of which code for proteins conserved in mammals. However, several questions remain unanswered. Perhaps the most important is: how do the Tor proteins "sense" amino acid limitation? While it has been proposed (Iiboshi et al. 1999) that the sensing mechanism involves detection of uncharged tRNAs (akin to GCN2; reviewed in Hinnebusch 2000), this remains controversial. Furthermore, the relative specificity of leucine in signaling to 4E-BP1 and to S6K1 in mammalian cells suggests that a "leucine receptor" may be present. An understanding of signaling downstream of the Tor proteins is also incomplete. While several exciting candidates (such as the PP2A-related phosphatases and Tap42p) have been postulated to mediate the rapamycin effects, the link between these proteins and downstream targets remains obscure. For FRAP/mTOR, the absence of a knock-out animal has limited the study of its function to rapamycin-inhibitable activities. It is likely that, similar to the yeast Tor2p, the mammalian FRAP/mTOR also exhibits rapamycin-resistant activities, which remain uncharacterized. Finally, while the Tor proteins have been proposed to serve as adapters or scaffolds, few Tor binding partners have been identified, and the roles of these interactions (as, for example, in the case of gephyrin) remain to be elucidated.

Acknowledgements. Work in the authors' laboratory was supported by grants from the National Cancer Institute of Canada and the Howard Hughes Medical Institute (HHMI). A.-C.G. is supported by a doctoral award from the Medical Research Council of Canada (MRC). N.S. is an MRC Distinguished Scientist and an HHMI International Scholar.

References

Abraham RT, Wiederrecht GJ (1996) Immunopharmacology of rapamycin. Annu Rev Immunol 14:483–510

Alarcon CM, Heitman J, Cardenas ME (1999) Protein kinase activity and identification of a toxic effector domain of the target of rapamycin TOR proteins in yeast. Mol Biol Cell 10:2531–2546

Albers MW, Williams RT, Brown EJ, Tanaka A, Hall FL, Schreiber SL (1993) FKBP-rapamycin inhibits a cyclin-dependent kinase activity and a cyclin D1-Cdk association in early G1 of an osteosarcoma cell line. J Biol Chem 268:22825–22829

Andrade MA, Bork P (1995) HEAT repeats in the Huntington's disease protein [letter]. Nat Genet 11:115–116

Baker H, Sidorowicz A, Sehgal SN, Vezina C (1978) Rapamycin (AY-22,989), a new antifungal antibiotic. III. In vitro and in vivo evaluation. J Antibiot (Tokyo) 31:539–545

Barbet NC, Schneider U, Helliwell SB, Stansfield I, Tuite MF, Hall MN (1996) TOR controls translation initiation and early G1 progression in yeast. Mol Biol Cell 7:25–42

Beck T, Hall MN (1999a) The TOR signalling pathway controls nuclear localization of nutrient-regulated transcription factors. Nature 402:689–692

Beck T, Schmidt A, Hall MN (1999b) Starvation induces vacuolar targeting and degradation of the tryptophan permease in yeast. J Cell Biol 146:1227–1238

Belsham GJ, Denton RM (1980) The effect of insulin and adrenaline on the phosphorylation of a 22 000-molecular weight protein within isolated fat cells; possible identification as the inhibitor-1 of the "general phosphatase". Biochem Soc Trans 8:382–383

Belsham GJ, Brownsey RW, Denton RM (1982) Reversibility of the insulin-stimulated phosphorylation of ATP citrate lyase and a cytoplasmic protein of subunit Mr 22000 in adipose tissue. Biochem J 204:345–352

Beretta L, Gingras A-C, Svitkin YV, Hall MN, Sonenberg N (1996) Rapamycin blocks the phosphorylation of 4E-BP1 and inhibits cap-dependent initiation of translation. EMBO J 15:658–664

Berset C, Trachsel H, Altmann M (1998) The TOR (target of rapamycin) signal transduction pathway regulates the stability of translation initiation factor eIF4G in the yeast *Saccharomyces cerevisiae*. Proc Natl Acad Sci USA 95:4264–4269

Bierer BE, Mattila PS, Standaert RF, Herzenberg LA, Burakoff SJ, Crabtree G, Schreiber SL (1990) Two distinct signal transmission pathways in T lymphocytes are inhibited by complexes formed between an immunophilin and either FK506 or rapamycin. Proc Natl Acad Sci USA 87:9231–9235

Bierer BE, Schreiber SL, Burakoff SJ (1991) The effect of the immunosuppressant FK-506 on alternate pathways of T cell activation. Eur J Immunol 21:439–445

Blackshear PJ, Nemenoff RA, Avruch J (1982) Preliminary characterization of a heat-stable protein from rat adipose tissue whose phosphorylation is stimulated by insulin. Biochem J 204:817–824

Blackshear PJ, Nemenoff RA, Avruch J (1983) Insulin and growth factors stimulate the phosphorylation of a Mr-22000 protein in 3T3-L1 adipocytes. Biochem J 214:11–19

Bosotti R, Isacchi A, Sonnhammer EL (2000) FAT: a novel domain in PIK-related kinases. Trends Biochem Sci 25:225–227

Bradley D (1999) FDA approves new immunosuppressants. Pharmaceutical Sci Tech Today 2:472

Brown EJ, Schreiber SL (1996) A signaling pathway to translational control. Cell 86:517–520

Brown EJ, Albers MW, Shin TB, Ichikawa K, Keith CT, Lane WS, Schreiber SL (1994) A mammalian protein targeted by G1-arresting rapamycin-receptor complex. Nature 369:756–758

Brown EJ, Beal PA, Keith CT, Chen J, Shin TB, Schreiber SL (1995) Control of p70 s6 kinase by kinase activity of FRAP in vivo. Nature 377:441–446

Brunn GJ, Williams J, Sabers C, Wiederrecht G, Lawrence JC Jr, Abraham RT (1996) Direct inhibition of the signaling functions of the mammalian target of rapamycin by the phosphoinositide 3-kinase inhibitors, wortmannin and LY294002. EMBO J 15:5256–5267

Brunn GJ, Fadden P, Haystead TAJ, Lawrence JC Jr (1997a) The mammalian target of rapamycin phosphorylates sites having a (Ser/Thr)-Pro motif and is activated by antibodies to a region near its COOH terminus. J Biol Chem 272:32547–32550

Brunn GJ, Hudson CC, Sekulic A, Williams JM, Hosoi H, Houghton PJ, Lawrence JC Jr, Abraham RT (1997b) Phosphorylation of the translational repressor PHAS-I by the mammalian target of rapamycin. Science 277:99–101

Burnett PE, Barrow RK, Cohen NA, Snyder SH, Sabatini DM (1998) RAFT1 phosphorylation of the translational regulators p70 S6 kinase and 4E-BP1. Proc Natl Acad Sci USA 95:1432–1437

Cafferkey R, Young PR, McLaughlin MM, Bergsma DJ, Koltin Y, Sathe GM, Faucette L, Eng WK, Johnson RK, Livi GP (1993) Dominant missense mutations in a novel yeast protein related to mammalian phosphatidylinositol 3-kinase and VPS34 abrogate rapamycin cytotoxicity. Mol Cell Biol 13:6012–6023

Cafferkey R, McLaughlin MM, Young PR, Johnson RK, Livi GP (1994) Yeast TOR (DRR) proteins: amino-acid sequence alignment and identification of structural motifs. Gene 141:133–136

Campbell LE, Wang X, Proud CG (1999) Nutrients differentially regulate multiple translation factors and their control by insulin. Biochem J 344:433–441

Canman CE, Lim DS (1998) The role of ATM in DNA damage responses and cancer. Oncogene 17:3301–3308

Cardenas ME, Heitman J (1995) FKBP12-rapamycin target TOR2 is a vacuolar protein with an associated phosphatidylinositol-4 kinase activity. EMBO J 14:5892–5907

Cardenas ME, Cutler NS, Lorenz MC, Di Como CJ, Heitman J (1999) The TOR signaling cascade regulates gene expression in response to nutrients. Genes Dev 13:3271–3279

Chen Y, Chen H, Rhoad AE, Warner L, Caggiano TJ, Failli A, Zhang H, Hsiao CL, Nakanishi K, Molnar-Kimber KL (1994) A putative sirolimus (rapamycin) effector protein. Biochem Biophys Res Commun 203:1–7

Chen J, Zheng XF, Brown EJ, Schreiber SL (1995) Identification of an 11-kDa FKBP12-rapamycin-binding domain within the 289-kDa FKBP12-rapamycin-associated protein and characterization of a critical serine residue. Proc Natl Acad Sci USA 92:4947–4951

Chen J, Peterson RT, Schreiber SL (1998) Alpha 4 associates with protein phosphatases 2 A, 4, and 6. Biochem Biophys Res Commun 247:827–832

Chiu MI, Katz H, Berlin V (1994) RAPT1, a mammalian homolog of yeast Tor, interacts with the FKBP12/rapamycin complex. Proc Natl Acad Sci USA 91:12574–12578

Choi J, Chen J, Schreiber SL, Clardy J (1996) Structure of the FKBP12-rapamycin complex interacting with the binding domain of human FRAP. Science 273:239–242

Cosentino GP, Schmelzle T, Haghighat A, Helliwell SB, Hall MN, Sonenberg N (2000) Eap1p, a novel eukaryotic translation initiation factor 4E-associated protein in *Saccharomyces cerevisiae*. Mol Cell Biol 20:4604–4613

Critchlow SE, Jackson SP (1998) DNA end-joining: from yeast to man. Trends Biochem Sci 23:394–398

Danaie P, Altmann M, Hall MN, Trachsel H, Helliwell SB (1999) CLN3 expression is sufficient to restore G1-to-S-phase progression in *Saccharomyces cerevisiae* mutants defective in translation initiation factor eIF4E. Biochem J 340 (Pt 1):135–141

De Moor CH, Jansen M, Bonte EJ, Thomas AA, Sussenbach JS, Van den Brande JL (1994) Influence of the four leader sequences of the human insulin-like-growth-factor-2 mRNAs on the expression of reporter genes. Eur J Biochem 226:1039–1047

De Moor CH, Jansen M, Bonte EJ, Thomas AA, Sussenbach JS, Van Den Brande JL (1995) Proteins binding to the leader of the 6.0 kb mRNA of human insulin-like growth factor 2 influence translation. Biochem J 307 (Pt 1):225–231

Di Como CJ, Arndt KT (1996) Nutrients, via the Tor proteins, stimulate the association of Tap42 with type 2 A phosphatases. Genes Dev 10:1904–1916

Dolinski K, Muir S, Cardenas M, Heitman J (1997) All cyclophilins and FK506 binding proteins are, individually and collectively, dispensable for viability in *Saccharomyces cerevisiae.* Proc Natl Acad Sci USA 94:13093–13098

Dufner A, Andjelkovic M, Burgering BM, Hemmings BA, Thomas G (1999) Protein kinase B localization and activation differentially affect S6 kinase 1 activity and eukaryotic translation initiation factor 4E-binding protein 1 phosphorylation. Mol Cell Biol 19:4525–4534

Dumont FJ, Melino MR, Staruch MJ, Koprak SL, Fischer PA, Sigal NH (1990a) The immunosuppressive macrolides FK-506 and rapamycin act as reciprocal antagonists in murine T cells. J Immunol 144:1418–1424

Dumont FJ, Staruch MJ, Koprak SL, Melino MR, Sigal NH (1990b) Distinct mechanisms of suppression of murine T cell activation by the related macrolides FK-506 and rapamycin. J Immunol 144:251–258

Dumont FJ, Staruch MJ, Grammer T, Blenis J, Kastner CA, Rupprecht KM (1995) Dominant mutations confer resistance to the immunosuppressant, rapamycin, in variants of a T cell lymphoma. Cell Immunol 163:70–79

Eng CP, Sehgal SN, Vezina C (1984) Activity of rapamycin (AY-22,989) against transplanted tumors. J Antibiot (Tokyo) 37:1231–1237

Fadden P, Haystead TA, Lawrence JC Jr (1997) Identification of phosphorylation sites in the translational regulator, PHAS-I, that are controlled by insulin and rapamycin in rat adipocytes. J Biol Chem 272:10240–10247

Featherstone C, Jackson SP (1999) DNA double-strand break repair. Curr Biol 9:R759–761

Fumagalli S, Thomas G (2000) S6 phosphorylation and signal transduction. In: Sonenberg N, Hershey JWB, Mathews MB (eds) Translational control of gene expression. Cold Spring Harbor Laboratory Press, Plainview, NY, pp 695–718

Gallego C, Gari E, Colomina N, Herrero E, Aldea M (1997) The Cln3 cyclin is down-regulated by translational repression and degradation during the G1 arrest caused by nitrogen deprivation in budding yeast. EMBO J 16:7196–7206

Geballe AP, Sachs MS (2000) Translational control by upstream open reading frames. In: Sonenberg N, Hershey JWB, Mathews MB (eds) Translational control of gene expression. Cold Spring Harbor Laboratory Press, Plainview, NY, pp 595–614

Gingras A-C, Kennedy SG, O'Leary MA, Sonenberg N, Hay N (1998) 4E-BP1, a repressor of mRNA translation, is phosphorylated and inactivated by the Akt(PKB) signaling pathway. Genes Dev 12:502–513

Gingras A-C, Raught B, Sonenberg N (1999a) eIF4 Initiation Factors: Effectors of mRNA recruitment to ribosomes and regulators of translation. Annu Rev Biochem 68:913–963

Gingras A-C, Gygi SP, Raught B, Polakiewicz RD, Abraham RT, Hoekstra MF, Aebersold R, Sonenberg N (1999b) Regulation of 4E-BP1 phosphorylation: a novel two-step mechanism. Genes Dev 13:1422–1437

Gingras A-C, Raught B, Gygi SP, Polakiewicz RD, Marcotrigiano J, Miron M, Poulin F, Burley SK, Aebersold R, Sonenberg N (2001) Serum-sensitive phosphorylation of a subset of sites in the translational inhibitor 4E-BP1 abrogates its binding to eIF4E. (in preparation)

Gothel SF, Marahiel MA (1999) Peptidyl-prolyl *cis-trans* isomerases, a superfamily of ubiquitous folding catalysts. Cell Mol Life Sci 55:423–436

Graves LM, Bornfeldt KE, Argast GM, Krebs EG, Kong X, Lin TA, Lawrence JC Jr (1995) cAMP- and rapamycin-sensitive regulation of the association of eukaryotic initiation factor 4E and the translational regulator PHAS-I in aortic smooth muscle cells. Proc Natl Acad Sci USA 92:7222–7226

Groves MR, Barford D (1999) Topological characteristics of helical repeat proteins. Curr Opin Struct Biol 9:383–389

Groves MR, Hanlon N, Turowski P, Hemmings BA, Barford D (1999) The structure of the protein phosphatase 2 A PR65/A subunit reveals the conformation of its 15 tandemly repeated HEAT motifs. Cell 96:99–110

Haendler B, Keller R, Hiestand PC, Kocher HP, Wegmann G, Movva NR (1989) Yeast cyclophilin: isolation and characterization of the protein, cDNA and gene. Gene 83:39–46

Hara K, Yonezawa K, Kozlowski MT, Sugimoto T, Andrabi K, Weng QP, Kasuga M, Nishimoto I, Avruch J (1997) Regulation of eIF-4E BP1 phosphorylation by mTOR. J Biol Chem 272: 26457–26463

Hara K, Yonezawa K, Weng QP, Kozlowski MT, Belham C, Avruch J (1998) Amino acid sufficiency and mTOR regulate p70 S6 kinase and eIF-4E BP1 through a common effector mechanism. J Biol Chem 273:14484–14494

Harding MW, Galat A, Uehling DE, Schreiber SL (1989) A receptor for the immunosuppressant FK506 is a *cis-trans* peptidyl-prolyl isomerase. Nature 341:758–760

Hardwick JS, Kuruvilla FG, Tong JK, Shamji AF, Schreiber SL (1999) Rapamycin-modulated transcription defines the subset of nutrient-sensitive signaling pathways directly controlled by the Tor proteins. Proc Natl Acad Sci USA 96:14866–14870

Hartley KO, Gell D, Smith GC, Zhang H, Divecha N, Connelly MA, Admon A, Lees-Miller SP, Anderson CW, Jackson SP (1995) DNA-dependent protein kinase catalytic subunit: a relative of phosphatidylinositol 3-kinase and the ataxia telangiectasia gene product. Cell 82:849–856

Heesom KJ, Avison MB, Diggle TA, Denton RM (1998) Insulin-stimulated kinase from rat fat cells that phosphorylates initiation factor-4E binding protein 1 on the rapamycin-insensitive site (serine-111). Biochem J 336:39–48

Heesom KJ, Denton RM (1999) Dissociation of the eukaryotic initiation factor-4E/4E-BP1 complex involves phosphorylation of 4E-BP1 by an mTOR-associated kinase. FEBS Lett 457:489–493

Heitman J, Movva NR, Hall MN (1991a) Targets for cell cycle arrest by the immunosuppressant rapamycin in yeast. Science 253:905–909

Heitman J, Movva NR, Hiestand PC, Hall MN (1991b) FK 506-binding protein proline rotamase is a target for the immunosuppressive agent FK 506 in *Saccharomyces cerevisiae*. Proc Natl Acad Sci USA 88:1948–1952

Helliwell SB, Wagner P, Kunz J, Deuter-Reinhard M, Henriquez R, Hall MN (1994) TOR1 and TOR2 are structurally and functionally similar but not identical phosphatidylinositol kinase homologues in yeast. Mol Biol Cell 5:105–118

Hershey JWB, Merrick WC (2000) Pathway and mechanism of initiation of protein synthesis. In: Sonenberg N, Hershey JWB, Mathews MB (eds) Translational control of gene expression. Cold Spring Harbor Laboratory Press, Plainview, NY, pp 33–88

Hinnebusch AG (2000) Mechanism and regulation of initiator methionyl-tRNA binding to ribosomes. In: Sonenberg N, Hershey JWB, Mathews MB (eds) Translational control of gene expression. Cold Spring Harbor Laboratory Press, Plainview, NY, pp 185–244

Hu C, Pang S, Kong X, Velleca M, Lawrence JC Jr (1994) Molecular cloning and tissue distribution of PHAS-I, an intracellular target for insulin and growth factors. Proc Natl Acad Sci USA 91:3730–3734

Iiboshi Y, Papst PJ, Kawasome H, Hosoi H, Abraham RT, Houghton PJ, Terada N (1999) Amino-acid-dependent control of p70s6k. J Biol Chem 274:1092–1099

Inui S, Sanjo H, Maeda K, Yamamoto H, Miyamoto E, Sakaguchi N (1998) Ig receptor binding protein 1 (α4) is associated with a rapamycin-sensitive signal transduction in lymphocytes through direct binding to the catalytic subunit of protein phosphatase 2 A. Blood 92:539–546

Jefferies HB, Thomas G (1996) Ribosomal protein S6 phosphorylation and signal transduction. In: Sonenberg N, Hershey JWB, Mathews MB (eds) Translational control of gene expression. Cold Spring Harbor Laboratory Press, Plainview, NY, pp 389–409

Jefferies HB, Reinhard C, Kozma SC, Thomas G (1994) Rapamycin selectively represses translation of the "polypyrimidine tract" mRNA family. Proc Natl Acad Sci USA 91:4441–4445

Jeggo PA, Carr AM, Lehmann AR (1998) Splitting the ATM: distinct repair and checkpoint defects in ataxia-telangiectasia. Trends Genet 14:312–316

Jiang Y, Broach JR (1999) Tor proteins and protein phosphatase 2 A reciprocally regulate Tap42 in controlling cell growth in yeast. EMBO J 18:2782–2792

Kay JE, Kromwel L, Doe SE, Denyer M (1991) Inhibition of T and B lymphocyte proliferation by rapamycin. Immunology 72:544–549

Kay JE, Smith MC, Frost V, Morgan GY (1996) Hypersensitivity to rapamycin of BJAB B lymphoblastoid cells. Immunology 87:390–395

Khaleghpour K, Pyronnet S, Gingras A-C, Sonenberg N (1999) Translational homeostasis: eukaryotic translation initiation factor 4E control of 4E-binding protein 1 and p70 S6 kinase activities. Mol Cell Biol 19:4302–4310

Kimball SR, Jefferson LS (2000) Regulation of translation initiation in mammalian cells by amino acids. In: Sonenberg N, Hershey JWB, Mathews MB (eds) Translational control of gene expression. Cold Spring Harbor Laboratory Press, Plainview, NY, pp 561–580

Kimball SR, Shantz LM, Horetsky RL, Jefferson LS (1999) Leucine regulates translation of specific mRNAs in L6 myoblasts through mTOR-mediated changes in availability of eIF4E and phosphorylation of ribosomal protein S6. J Biol Chem 274:11647–11652

Kino T, Hatanaka H, Hashimoto M, Nishiyama M, Goto T, Okuhara M, Kohsaka M, Aoki H, Imanaka H (1987a) FK-506, a novel immunosuppressant isolated from a Streptomyces. I. Fermentation, isolation, and physico-chemical and biological characteristics. J Antibiot (Tokyo) 40:1249–1255

Kino T, Hatanaka H, Miyata S, Inamura N, Nishiyama M, Yajima T, Goto T, Okuhara M, Kohsaka M, Aoki H et al. (1987b) FK-506, a novel immunosuppressant isolated from a *Streptomyces.* II. Immunosuppressive effect of FK-506 in vitro. J Antibiot (Tokyo) 40:1256–1265

Kohn AD, Barthel A, Kovacina KS, Boge A, Wallach B, Summers SA, Birnbaum MJ, Scott PH, Lawrence JC, Jr, Roth RA (1998) Construction and characterization of a conditionally active version of the serine/threonine kinase Akt. J Biol Chem 273:11937–11943

Koltin Y, Faucette L, Bergsma DJ, Levy MA, Cafferkey R, Koser PL, Johnson RK, Livi GP (1991) Rapamycin sensitivity in *Saccharomyces cerevisiae* is mediated by a peptidyl-prolyl cis-trans isomerase related to human FK506-binding protein. Mol Cell Biol 11:1718–1723

Kumar V, Pandey P, Sabatini D, Kumar M, Majumder PK, Bharti A, Carmichael G, Kufe D, Kharbanda S (2000a) Functional interaction between RAFT1/FRAP/mTOR and protein kinase cdelta in the regulation of cap-dependent initiation of translation. EMBO J 19:1087–1097

Kumar V, Sabatini D, Pandey P, Gingras A-C, Majumder PK, Kumar M, Yuan ZM, Carmichael G, Weichselbaum R, Sonenberg N, Kufe D, Kharbanda S (2000b) Regulation of the rapamycin and FKBP-target 1/mammalian target of rapamycin and cap-dependent initiation of translation by the c-Abl protein-tyrosine kinase. J Biol Chem 275:10779–10787

Kunz J, Henriquez R, Schneider U, Deuter-Reinhard M, Movva NR, Hall MN (1993) Target of rapamycin in yeast, TOR2, is an essential phosphatidylinositol kinase homolog required for G1 progression. Cell 73:585–596

Lavin MF, Shiloh Y (1997) The genetic defect in ataxia-telangiectasia. Annu Rev Immunol 15:177–202

Lazaris-Karatzas A, Montine KS, Sonenberg N (1990) Malignant transformation by a eukaryotic initiation factor subunit that binds to mRNA 5′ cap. Nature 345:544–547

Lin TA, Kong X, Haystead TA, Pause A, Belsham G, Sonenberg N, Lawrence JC Jr (1994) PHAS-I as a link between mitogen-activated protein kinase and translation initiation. Science 266:653–656

Lin TA, Lawrence JC Jr (1996) Control of the translational regulators PHAS-I and PHAS-II by insulin and cAMP in 3T3-L1 adipocytes. J Biol Chem 271:30199–30204

Lorenz MC, Heitman J (1995) TOR mutations confer rapamycin resistance by preventing interaction with FKBP12-rapamycin. J Biol Chem 270:27531–27537

Lu KP (2000) Phosphorylation-dependent prolyl isomerization: a novel cell cycle regulatory mechanism. Prog Cell Cycle Res 4:83–96

Luo H, Chen H, Daloze P, St-Louis G, Wu J (1993) Anti-CD28 antibody- and IL-4-induced human T cell proliferation is sensitive to rapamycin. Clin Exp Immunol 94:371–376

Marcotrigiano J, Gingras A-C, Sonenberg N, Burley SK (1999) Cap-dependent translation initiation in eukaryotes is regulated by a molecular mimic of eIF4G. Mol Cell 3:707–716

Martel RR, Klicius J, Galet S (1977) Inhibition of the immune response by rapamycin, a new antifungal antibiotic. Can J Physiol Pharmacol 55:48–51

Meyuhas O, Hornstein E (2000) Translational control of TOP mRNAs. In: Sonenberg N, Hershey JWB, Mathews MB (eds) Translational control of gene expression. Cold Spring Harbor Laboratory Press, Plainview, NY, pp 671–694

Meyuhas O, Avni D, Shama S (1996) Translational control of ribosomal protein mRNAs in eukaryotes. In: Sonenberg N, Hershey JWB, Mathews MB (eds) Translational control. Cold Spring Harbor Laboratory Press, Plainview, NY, pp 363–388

Murata K, Wu J, Brautigan DL (1997) B cell receptor-associated protein alpha4 displays rapamycin-sensitive binding directly to the catalytic subunit of protein phosphatase 2 A. Proc Natl Acad Sci USA 94:10624–10629

Nanahoshi M, Nishiuma T, Tsujishita Y, Hara K, Inui S, Sakaguchi N, Yonezawa K (1998) Regulation of protein phosphatase 2 A catalytic activity by alpha4 protein and its yeast homolog TAP42. Biochem Biophys Res Commun 251:52–526

Nanahoshi M, Tsujishita Y, Tokunaga C, Inui S, Sakaguchi N, Hara K, Yonezawa K (1999) Alpha4 protein as a common regulator of type 2A-related serine/threonine protein phosphatases. FEBS Lett 446:108–112

Navé BT, Ouwens M, Withers DJ, Alessi DR, Shepherd PR (1999) Mammalian target of rapamycin is a direct target for protein kinase B: identification of a convergence point for opposing effects of insulin and amino-acid deficiency on protein translation. Biochem J 344:427–431

Nielsen FC, Christiansen J (1995) Posttranscriptional regulation of insulin-like growth factor II mRNA. Scand J Clin Lab Invest Suppl 220:37–46

Nielsen FC, Ostergaard L, Nielsen J, Christiansen J (1995) Growth-dependent translation of IGF-II mRNA by a rapamycin-sensitive pathway. Nature 377:358–362

Nielsen J, Christiansen J, Lykke-Andersen J, Johnsen AH, Wewer UM, Nielsen FC (1999) A family of insulin-like growth factor II mRNA-binding proteins represses translation in late development. Mol Cell Biol 19:1262–1270

Noda T, Ohsumi Y (1998) Tor, a phosphatidylinositol kinase homologue, controls autophagy in yeast. J Biol Chem 273:3963–3966

Patti ME, Brambilla E, Luzi L, Landaker EJ, Kahn CR (1998) Bidirectional modulation of insulin action by amino acids. J Clin Invest 101:1519–1529

Pause A, Belsham GJ, Gingras A-C, Donzé O, Lin TA, Lawrence JC Jr, Sonenberg N (1994) Insulin-dependent stimulation of protein synthesis by phosphorylation of a regulator of 5′-cap function. Nature 371:762–767

Pedersen S, Celis JE, Nielsen J, Christiansen J, Nielsen FC (1997) Distinct repression of translation by wortmannin and rapamycin. Eur J Biochem 247:449–456

Peterson RT, Desai BN, Hardwick JS, Schreiber SL (1999) Protein phosphatase 2 A interacts with the 70-kDa S6 kinase and is inactivated by inhibition of FKBP12-rapamycin-associated protein. Proc Natl Acad Sci USA 96:4438–4442

Peterson RT, Beal PA, Comb MJ, Schreiber SL (2000) FKBP12-rapamycin-associated protein (FRAP) autophosphorylates at serine 2481 under translationally repressive conditions. J Biol Chem 275:7416–7423

Polymenis M, Schmidt EV (1997) Coupling of cell division to cell growth by translational control of the G1 cyclin CLN3 in yeast. Genes Dev 11:2522–2531

Poulin F, Gingras A-C, Olsen H, Chevalier S, Sonenberg N (1998) 4E-BP3, a new member of the eukaryotic initiation factor 4E-binding protein family. J Biol Chem 273:14002–14007

Powers T, Walter P (1999) Regulation of ribosome biogenesis by the rapamycin-sensitive TOR-signaling pathway in *Saccharomyces cerevisiae.* Mol Biol Cell 10:987–1000

Powis G, Bonjouklian R, Berggren MM, Gallegos A, Abraham R, Ashendel C, Zalkow L, Matter WF, Dodge J, Grindey G (1994) Wortmannin, a potent and selective inhibitor of phosphatidylinositol-3-kinase. Cancer Res 54:2419–2423

Raught B, Gingras A-C, James A, Medina D, Sonenberg N, Rosen JM (1996) Expression of a translationally regulated, dominant-negative CCAAT/enhancer-binding protein beta isoform and up-regulation of the eukaryotic translation initiation factor 2/alpha are correlated with neoplastic transformation of mammary epithelial cells. Cancer Res 56:4382–4386

Raught B, Gingras A-C, Sonenberg N (2000a) Regulation of ribosomal recruitment in eukaryotes. In: Sonenberg N, Hershey JWB, Mathews MB (eds) Translational control of gene expression. Cold Spring Harbor Laboratory Press, Plainview, NY, pp 245–294

Raught B, Gingras A-C, Gygi SP, Imataka H, Morino S, Gradi A, Aebersold R, Sonenberg N (2000b) Serum-stimulated, rapamycin-sensitive phosphorylation sites in the eukaryotic translation initiation factor 4GI. EMBO J 19:434–444

Rosenwald IB, Chen JJ, Wang S, Savas L, London IM, Pullman J (1999) Upregulation of protein synthesis initiation factor eIF-4E is an early event during colon carcinogenesis. Oncogene 18:2507–2517

Rotman G, Shiloh Y (1998) ATM: from gene to function. Hum Mol Genet 7:1555–1563

Rotman G, Shiloh Y (1999) ATM: a mediator of multiple responses to genotoxic stress. Oncogene 18:6135–6144

Sabatini DM, Erdjument-Bromage H, Lui M, Tempst P, Snyder SH (1994) RAFT1: a mammalian protein that binds to FKBP12 in a rapamycin-dependent fashion and is homologous to yeast TORs. Cell 78:35–43

Sabatini DM, Barrow RK, Blackshaw S, Burnett PE, Lai MM, Field ME, Bahr BA, Kirsch J, Betz H, Snyder SH (1999) Interaction of RAFT1 with gephyrin required for rapamycin-sensitive signaling. Science 284:1161–1164

Sabers CJ, Martin MM, Brunn GJ, Williams JM, Dumont FJ, Wiederrecht G, Abraham RT (1995) Isolation of a protein target of the FKBP12-rapamycin complex in mammalian cells. J Biol Chem 270:815–822

Sarkaria JN, Tibbetts RS, Busby EC, Kennedy AP, Hill DE, Abraham RT (1998) Inhibition of phosphoinositide 3-kinase related kinases by the radiosensitizing agent wortmannin. Cancer Res 58:4375–4382

Schmidt A, Kunz J, Hall MN (1996) TOR2 is required for organization of the actin cytoskeleton in yeast. Proc Natl Acad Sci USA 93:13780–13785

Schmidt A, Bickle M, Beck T, Hall MN (1997) The yeast phosphatidylinositol kinase homolog TOR2 activates RHO1 and RHO2 via the exchange factor ROM2. Cell 88:531–542

Schmidt A, Beck T, Koller A, Kunz J, Hall MN (1998) The TOR nutrient signalling pathway phosphorylates NPR1 and inhibits turnover of the tryptophan permease. EMBO J 17:6924–6931

Scott PH, Lawrence JC Jr (1998) Attenuation of mammalian target of rapamycin activity by increased cAMP in 3T3-L1 cells. J Biol Chem 272:34496–34501

Scott PH, Brunn GJ, Kohn AD, Roth RA, Lawrence JC Jr (1998) Evidence of insulin-stimulated phosphorylation and activation of the mammalian target of rapamycin mediated by a protein kinase B signaling pathway. Proc Natl Acad Sci USA 95:7772–7777

Sehgal SN, Baker H, Vezina C (1975) Rapamycin (AY-22,989), a new antifungal antibiotic. II. Fermentation, isolation and characterization. J Antibiot (Tokyo) 28:727–732

Sekulic A, Hudson CC, Homme JL, Yin P, Otterness DM, Karnitz LM, Abraham RT (2000) A direct linkage between the phosphoinositide 3-Kinase-AKT signaling pathway and the mammalian target of rapamycin. Cancer Res 60:3504–3513

Shigemitsu K, Tsujishita Y, Hara K, Nanahoshi M, Avruch J, Yonezawa K (1999) Regulation of translational effectors by amino acid and mammalian target of rapamycin signaling pathways. Possible involvement of autophagy in cultured hepatoma cells. J Biol Chem 274:1058–1065

Shiloh Y (1997) Ataxia-telangiectasia and the Nijmegen breakage syndrome: related disorders but genes apart. Annu Rev Genet 31:635–662

Siekierka JJ, Hung SH, Poe M, Lin CS, Sigal NH (1989) A cytosolic binding protein for the immunosuppressant FK506 has peptidyl-prolyl isomerase activity but is distinct from cyclophilin. Nature 341:755–757

Siekierka JJ, Wiederrecht G, Greulich H, Boulton D, Hung SH, Cryan J, Hodges PJ, Sigal NH (1990) The cytosolic-binding protein for the immunosuppressant FK-506 is both a ubiquitous and highly conserved peptidyl-prolyl cis-trans isomerase. J Biol Chem 265:21011–21015

Singh K, Sun S, Vezina C (1979) Rapamycin (AY-22,989), a new antifungal antibiotic. IV. Mechanism of action. J Antibiot (Tokyo) 32:630–645

Smith GC, Jackson SP (1999) The DNA-dependent protein kinase. Genes Dev 13:916–934

Smith GC, Divecha N, Lakin ND, Jackson SP (1999) DNA-dependent protein kinase and related proteins. Biochem Soc Symp 64:91–104

Stan R, McLaughlin MM, Cafferkey R, Johnson RK, Rosenberg M, Livi GP (1994) Interaction between FKBP12-rapamycin and TOR involves a conserved serine residue. J Biol Chem 269:32027–32030

Sykes K, Gething MJ, Sambrook J (1993) Proline isomerases function during heat shock. Proc Natl Acad Sci USA 90:5853–5857

Takata M, Ogawa W, Kitamura T, Hino Y, Kuroda S, Kotani K, Klip A, Gingras A-C, Sonenberg N, Kasuga M (1999) Requirement for Akt (protein kinase B) in insulin-induced activation of glycogen synthase and phosphorylation of 4E-BP1 (PHAS-1). J Biol Chem 274:20611–20618

Tropschug M, Barthelmess IB, Neupert W (1989) Sensitivity to cyclosporin A is mediated by cyclophilin in *Neurospora crassa* and *Saccharomyces cerevisiae*. Nature 342:953–955

Ui M, Okada T, Hazeki K, Hazeki O (1995) Wortmannin as a unique probe for an intracellular signalling protein, phosphoinositide 3-kinase. Trends Biochem Sci 20:303–307

Vezina C, Kudelski A, Sehgal SN (1975) Rapamycin (AY-22,989), a new antifungal antibiotic. I. Taxonomy of the producing streptomycete and isolation of the active principle. J Antibiot (Tokyo) 28:721–726

Vilella-Bach M, Nuzzi P, Fang Y, Chen J (1999) The FKBP12-rapamycin-binding domain is required for FKBP12-rapamycin-associated protein kinase activity and G1 progression. J Biol Chem 274:4266–4272

Vlahos CJ, Matter WF, Hui KY, Brown RF (1994) A specific inhibitor of phosphatidylinositol 3-kinase, 2-(4morpholinyl)-8-phenyl-4H-1-benzopyran-4-one (LY294002). J Biol Chem 2169:5241–5248

von Manteuffel SR, Gingras A-C, Ming XF, Sonenberg N, Thomas G (1996) 4E-BP1 phosphorylation is mediated by the FRAP-p70s6k pathway and is independent of mitogen-activated protein kinase. Proc Natl Acad Sci USA 93:4076–4080

Wang X, Campbell LE, Miller CM, Proud CG (1998) Amino acid availability regulates p70 S6 kinase and multiple translation factors. Biochem J 334:261–267

Wiederrecht G, Brizuela L, Elliston K, Sigal NH, Siekierka JJ (1991) FKB1 encodes a nonessential FK 506-binding protein in *Saccharomyces cerevisiae* and contains regions suggesting homology to the cyclophilins. Proc Natl Acad Sci USA 88:1029–1033

Wymann MP, Bulgarelli-Leva G, Zvelebil MJ, Pirola L, Vanhaesebroeck B, Waterfield MD, Panayotou G (1996) Wortmannin inactivates phosphoinositide 3-kinase by covalent modification of Lys-802, a residue involved in the phosphate transfer reaction. Mol Cell Biol 16:1722–1733

Xu G, Kwon G, Marshall CA, Lin TA, Lawrence JC Jr, McDaniel ML (1998a) Branched-chain amino acids are essential in the regulation of PHAS-I and p70 S6 kinase by pancreatic beta-cells. A possible role in protein translation and mitogenic signaling. J Biol Chem 273:28178–28184

Xu G, Marshall CA, Lin TA, Kwon G, Munivenkatappa RB, Hill JR, Lawrence JC Jr, McDaniel ML (1998b) Insulin mediates glucose-stimulated phosphorylation of PHAS-I by pancreatic beta cells. An insulin-receptor mechanism for autoregulation of protein synthesis by translation. J Biol Chem 273:4485–4491

Zaragoza D, Ghavidel A, Heitman J, Schultz MC (1998) Rapamycin induces the G(0) program of transcriptional repression in yeast by interfering with the TOR signaling pathway. Mol Cell Biol 18:4463–4470

Zheng XF, Florentino D, Chen J, Crabtree GR, Schreiber SL (1995) TOR kinase domains are required for two distinct functions, only one of which is inhibited by rapamycin. Cell 82:121–130

Subject Index

Printing (Computer to Film): Saladruck, Berlin
Binding: Stürtz AG, Würzburg